A Really Inconvenient Truth

The Case Against the Theory of Anthropogenic Global Warming

By Philip M. Fishman
Foreword by Col. Ronald D. Harris M.D.

ACKNOWLEDGEMENTS

To my good friend from our early days in the U.S. Army Chemical Corps, Dr. John N. Hubbell, retired physics professor from Louisiana State University for reviewing and offering constructive comments on chapter 12 statistical validity.

To my new friend and colleague, Colonel Ronald D. Harris, M. D. for his review of my manuscript and helpful comments as well as writing the foreword.

To Elaine Lanmon of Just Ink Graphic Design for the format and design of this book as well as her helpful advice.

I would be remiss if I did not also mention the many great teachers and professors that set me on the road, sometimes in spite of myself. There are far too many to mention them all and some names I have frankly forgotten, but these few stand out in my memory: Roy Long, high school math; Charles McClary, high school chemistry; Warren Pemberton, high school physics; and Dr. Frederick Schmidt, college chemistry.

And last but not least, to my darling wife, Sara, for her continual encouragement throughout our years of marriage.

Please check out my other books
The others are all in the Jefferson County
Library system.
    Google "Hoover Swan" Phil Fisher
(The young lady in the photo is our daughter,
Missy.

Foreword

"The Arctic ocean is warming up, icebergs are growing scarcer, and in some places seals are finding waters too hot ..." "...Reports from fishermen, seal hunters and explorers...all point to a radical change in climactic conditions and hitherto unheard- of temperatures in the Arctic zone..."

So what's the big deal with this excerpt from an Associated Press article that seems to be so common these days? Not a thing except that this article appeared in a Washington Post edition dated November 2, 1922. Ninety years ago the same "sky is falling" prophesy was being touted. Those who subscribe to the global warming theory may well say that the warming trend actually began in the mid- eighteen hundreds and the above report was a continuation of the trend up to the present. Except in the interim there was a period in the 1970s when Time Magazine ran stories that claimed the Earth was entering another "Ice Age"!

I positively enjoyed reading this book. It is educational but entertaining as well, spiced with wry wit throughout. I found that it is an honest representation of the facts surrounding the issues of climate change and alleged human caused global warming, based on unscientific hyperbole.

I feel my background qualifies me to make that assessment, considering my active involvement in the world of a science over a 49-year career. During that career I have conducted and overseen bench research, animal research, and a large number of clinical research projects. My career has ranged from the private practice of medicine to the academic world as a full professor at major universities to the military as senior medical officer in the Department of Defense vaccine agency. As a university scholar and researcher, I have developed a firm understanding and appreciation of the scientific method, and have also seen my fair share of sham theories.

In this book, Fishman not only delivers a solid and fully documented narrative, but takes pains to adhere to an indisputable scientific method as he describes the various aspects of climate. I recognize that climate change has been a contentious issue for many years and have followed arguments on the pros and cons very closely. I think that some who are convinced that the Earth is warming and that something must be done to stop it are doing it with the best of intentions; however, I find the majority of these arguments to be fallacious and lacking a scientific basis. Establishing laws that would punish industries and in turn damage the economy because of a political agenda is simply wrong.

I had the pleasure of getting to know Phil Fishman through a very good friend of mine who is a retired an Air Force colonel. When I heard about Phil's book I was delighted to play a role, since I admire his tenacity and perseverance in sifting through the enormous amount of information and misinformation available on a subject that has garnered as much mystical controversy as climate has in the past several years.

Phil is a scientist. His science education started with a degree in chemistry. Following graduation he was commissioned a second lieutenant in the U.S. Army Chemical Corps where he first served as a platoon leader of a chemical supply unit and later as the executive officer of a technical intelligence unit. After being honorably discharged from the Army he worked in various facets of the chemical industry for over 40 years. Upon retirement he went back to school to obtain certification and began teaching science to schoolchildren. Due to his science background and knowledge of the scientific method, he became intrigued by the growing debate concerning global warming. After researching and reviewing the available data and facts on the matter, his intrigue turned to concern as he realized that there was a considerable amount of intentional misinformation being disseminated in order to create an apparent mystique. Further, when anyone challenged the doctrine they were being denigrated. He embarked in an effort to become further educated in the subject, compiling facts and researching the science for more than five years. This book is the culmination of his research and findings.

Part I of the book, titled "inconvenient science" takes you through an educational process as Fishman discusses various aspects of climatology, beginning with a chapter on "the scientific method" and why it is vital to be able to differentiate actual knowledge from belief.

The second chapter explains the rationale of the theory of human caused global warming. This view is dubbed the "The Theory," and its followers are referred to as "theorists". The third chapter, which is of critical importance discusses the methodology used in measuring the Earth's temperature and the accuracy of the temperature recordings; and chapter six discusses the methodology used in determining global temperatures from ages past when measuring instruments had not yet been invented . One fact that is quite interesting to me, and is widely unknown, is that the temperatures of the Earth were warmer in medieval times than they are today. Theorists have tried to hide that fact since it refutes the main premise of the theory, i.e., that higher temperatures today are a result of higher carbon dioxide levels in the atmosphere. Over the ages, both prior to the appearance of Man and since, the Earth has gone from a steaming heated tropical environment to ice ages, and then back. Humans certainly cannot be blamed for affecting climate before they were here; and their role in influencing climate after their arrival appears minimal at most.

Chapter four deals with "greenhouse gases" and discusses why that designation is a misnomer that conveys an erroneous view. Fishman further points out that carbon dioxide is probably one of the least important contributors to warming and that water vapor, surprisingly, has a tremendously stronger effect. Amazingly the major role played by the sun and solar activity affecting the Earth's climate is almost completely ignored by the theorists.

I was entertained, as well as enlightened, with Phil Fishman's observations on several of the corollaries of the "theory". One of its most dramatic, yet shamelessly flawed, symbols has been the plight of the polar bear. Theorists would have us believe this great and powerful animal is being decimated at the hands of human-caused warming. In fact, however, the most recent surveys indicate that not only have polar bears survived some temporary (and historically trending) recession of the Arctic sea ice, but also their numbers have more recently been shown to have significantly increased.

Part II shows that while many statistics are utilized to support the theorists' views, statistics can be, and have been, skillfully manipulated to tell a misleading story. This section points out how these manipulated statistics have been used to mislead and are neither a scientific, nor truthful basis for their arguments.

Part III includes information on economics related to the climate change arguments. "Alternative energy sources" are investigated and are exposed as not providing the panacea that theorists would wish for. One example of a really absurd economic alternative energy source is the use of ethanol as a replacement for gasoline burning engines.

Not only does the production of suitable ethanol (low water content) require more energy than is derived from the burning of ethanol, but it has resulted in the deleterious, and carbon increasing, effect of razing millions of acres of rain forest in order to farm and fertilize cornfields.

Part IV deals with "inconvenient facts" and shows how facts themselves may be ignored, or manipulated, or covered-up, for the purpose of deceiving the public. One fact that theorists like to ignore, but can't, is the lack of warming since 1998. The concomitant reduction in solar activity over this time period still seems to escape recognition by the theorists and the lack of warming is dismissed, however, as a short term aberration. Phil wryly asks how long an aberration must persist until it is no longer an aberration.

The concluding section discusses "inconvenient uncertainties", including the uncertainties of measurement, limits of our knowledge, and the difficulties in predicting the future.

Phil doesn't brow beat people into one way of thinking versus another. He presents the arguments for and against and then asks the readers to come to their own conclusion, now having information available to them about both sides of the issue. While we most certainly all want the best possible environment for ourselves, our children and grandchildren, the methods of getting where we need to be should be made from a scientific basis and not a political or mercenary agenda. People need to inform themselves on such an important issue to take the power away from those who would misuse science for their own ends.

Colonel Ronald D. Harris, M.D., USAF, FACR, FSU
La Jolla, CA

Preface

"So profound is our ignorance, and so high our presumption, that we marvel when we hear of the extinction of an organic being <or other calamities>; and as we do not see the cause, we invoke cataclysms to desolate the world, or invent laws on the duration of the forms of life!"
Charles Darwin – On the Origin of Species

My apologies to the Father of Modern Biology and his descendants for my insert into his quotation. I obviously cannot be certain that he would agree with the thought, but I have reason to believe that he would.

TABLE OF CONTENTS

INTRODUCTION

"It ain't what you don't know that gets you into trouble. It's what you know for sure that just ain't so."

The above quotation from Mark Twain was used by Nobel Laureate and ex- Vice President Al Gore at the beginning of his treatise on global warming, titled, An Inconvenient Truth. That Gore would have used this particular quotation, strikes me as ironic, considering that he and those who accept the theory as Gospel are so sure. He could not see that the admonition was directed at him as well as anyone absolutely certain of anything. Apparently, Mr. Gore's arrogance blinded him to the irony. He "knows it for sure" and after all the matter is "settled".

Since I dispute the statement that the matter is settled, I guess I would fall into the category of "denier", but the term has such a nasty connotation, I prefer the word, dissenter. Moreover, maybe we should reserve the term for the real deniers, who just may happen to be those who are using the term against us dissenters. I don't want to fall into Mark Twain's trap of knowing it for sure; so you may consider me someone who is fairly certain that the theory of anthropogenic (human caused) global warming, hereinafter referred to as The Theory, is bogus. You, the readers, will be the jury, so let us look at the evidence together before anyone jumps to conclusions.

A really inconvenient truth is that The Theory may be the largest and most brilliantly concocted hoax foisted on the world in all of history. There have been many hoaxes over the ages (centuries), but they have been (fairly) localized. The internet and global media, however, have totally altered the situation and now make it easy to spread the word, factual or not. The UN and governments around the world have bought into The Theory and are attempting to enact laws to combat the perceived catastrophe that is inevitable if we do not take drastic measures.

Adolph Hitler was a master of propaganda at persuading the masses. In his book, Mein Kampf, Hitler revealed his technique:

"By means of shrewd lies, unremittingly repeated, it is possible to make people believe that heaven is hell -- and hell heaven. The greater the lie, the more readily it will be believed."

What are the ingredients of a successful hoax?

    1. As preposterous as it may be, it must be believable.

2. In order to be believable, there must be some observable and easily established facts.

3. It must be supported by otherwise credible and respectable individuals. And

4. It must create a hysteria.

Let me be clear. This hoax is very different from the stereotype in that it did not start out as a hoax. It began with some very striking data, which when coupled with some known scientific facts and a historic fact, made for a plausible first hypothesis. Presumably, most of The Theory adherents, hereinafter referred to as theorists, truly believe (or more accurately, believe they know) that The Theory is a fact. The politicians, including Al Gore and lay people ignorant of the scientific method can be excused for their devotion. The scientists who have been at the center of the controversy, pushing The Theory, know better and should be ashamed of themselves for "cherry picking" favorable data, covering up contradictory data, and trying to intimidate and silence honest critics.

There is no governing board for scientists as there are for lawyers and doctors, which would disbar a lawyer or revoke a doctor's license for this kind of malfeasance, and for good reason. Scientists are encouraged to think "out of the box". This is where the great scientific discoveries have come from. When a crackpot comes along with some weird off- the-wall hypothesis, peer review will normally quickly resolve the issue. In this case, there was just enough science initially to make The Theory appear feasible. From there, after governments took an interest; it took on a life of its own. A second irony is the charge by the theorists that the only scientists who reject even any part of The Theory have been "bought" by the oil companies and others who profit from the burning of fossil fuels. The fact is that the big money by far is the government money, supporting "research" into global warming.

Do the documented misinformation, cover-up, and refusal to listen to the many reputable scientists, who challenge various assertions of The Theory in themselves, disprove The Theory? Of course not, but they certainly create a credibility gap when the theorists proclaim that the science is settled and there is no need for further discussion or inquiry.

Before we begin however, I would like to set the record straight on what has created a lot of confusion (which I believe from some quarters, was intentional). We dissenters do not dispute the fact that carbon dioxide is a greenhouse gas, or that direct readings taken from 1952 show a distinct upward trend. Most of us also agree that average global temperatures had been in a slight uptrend from perhaps the mid 1800's and more certainly from 1970 to about 1998. A smaller number may even agree that the increasing carbon dioxide levels are probably responsible for the slight uptrend of temperatures since the beginning of the industrial revolution. Where we depart adamantly from The Theory is the premise that temperatures have reached levels unseen for thousands or tens of thousands years and are in an accelerating uptrend, leading us to a global catastrophe if drastic measures to abate CO2 emissions are not immediately undertaken.

If you ask me if I believe the earth is warming, I would ask you to clarify the question.

If you mean "warming at this instant and continuing to warm into the immediate future", I would have to honestly say that I have no more insight into that than I would in trying to predict the direction of the stock market at any given time. The market will go up and it will go down. That, I can say with certainty. I just can't tell you when.

In Mr. Gore's book a casual reader might infer that he had formally studied climatology for many years before he went into politics. Statements such as these "…studied under Professor Roger Revelle", and "had been a student of climate for seventeen years before writing this book" might lead one to that conclusion. Because of privacy laws, I was unable to view Mr. Gore's transcript, but his biography states that his bachelor's degree from Harvard was in government. Later, he entered divinity school and then law school at Vanderbilt, but dropped out of both before graduating. I strongly suspect that the course Mr. Gore took under Professor Revelle was Gore's only college science course, unless one wants to include political science.

In fairness, I admit that I also am not a climatologist, but my education and career in chemistry have given me a firm understanding of the scientific method. I do not know if the scientific method was taught in Revelle's course, but if so, Mr. Gore seems to have forgotten it or dismissed it as unimportant.

What I would like to do in this book is to take an honest look at the science, statistics, economics, facts, and uncertainties behind the theory of anthropogenic warming and then permit you, the jury, to render your verdict. First let me address the proposition that planet earth is in a heating spell. It is interesting that as recently as forty years ago; there was a lot of hype among certain scientists, politicians, and reporters about the coming ice age. Even the theorists acknowledge the fact that over the eons there have been dramatic swings in global temperatures and that there have been periods a lot warmer (and cooler) than now. What they say is different now is that global temperatures have been going up at an alarming and accelerating rate over the last 150 years, which they relate primarily to industrialization and the emission of carbon dioxide.

We will address this issue in chapters two and three, but before we get into the meat of the argument, I would like to make a final point. With all the name calling and disparaging remarks aimed at us dissenters, although I have not seen it explicitly stated, there is the strong implication that any who deny the validity of The Theory, are indifferent to the future of the planet and unborn generations. I take strong offense, as I am sure all of my fellow dissenters do, of the suggestion that we care less about posterity than they do. In fact I believe a strong case can be made that it is just the opposite. The scientists who are theorists are the ones who apparently value their own livelihood more than the future since their manipulation of science is endangering posterity by damaging the credibility of the field that has brought so much good to so many people.

Part I
Inconvenient Science

"Scientists sometimes forget that the business of science is pursuing truth, not marketing their own version of truth."

The author

Chapter 1
The Scientific Method

Belief vs. Knowledge

What do we know? How do we know that it is true? It should be clear that observation alone doesn't insure accuracy. Several people having observed the same scene, when asked to describe what they saw, may have wildly different impressions. Just sit in a court room and listen to eye- witness accounts of a crime. You would be astounded at how different the accounts can be.

Psychology has taught us that perception is highly influenced by prior experience or conditioning. One can therefore understand how Mr. Gore became such a vehement Theorist. He acknowledges that his mother's repetitive readings of Silent Spring to him and his sister left a tremendous impression on him. It is obvious that that early indoctrination impaired his objectivity and tainted his perception towards all things concerned with the environment and industry.

We all believe, but we differ greatly in what we believe as well as the basis and certainty of those beliefs. Just the act of thinking involves belief. For example, when we look up at the moon, we see what appears to be two dimensional, but our mind registers a spherical object. In other words, we believe that the moon and, for that matter, all heavenly bodies are generally spherical. Columbus believed the world was spherical, but he didn't know it to be a fact. He had a strong hunch that it was so, based on observation and perception as he had watched ships slowly disappear over the horizon. But don't you know that his belief was not quite so strong after he had sailed over a month without seeing any trace of land?

If I hold three aces in the card game of five-card stud I should feel pretty confident of winning the hand. However, my belief that I have the best hand may change with the circumstances. If one of my aces is down (hidden from view), all anyone at the table knows is that I have two aces. If the best other hand at the table has only one pair showing after all five cards are dealt, I know that I have the winning hand. If, alternatively, there is someone with three of a kind showing, the odds are that I still have the winning hand. But I cannot be quite as certain in my belief since if that person's down card matches up with his odd card or his three of a kind he wins with either a "full house" or four of a kind. Now consider the possibility that all three of my aces are up and my opponent with, say three queens showing, has raised strongly against me after the fifth card has been dealt. The odds still favor my hand, but I now have to question why my opponent has raised if he knows I have the winning hand. Probably, he is just bluffing. I still believe I will win the pot but I am far less certain as I match the bet to call.

Until 1905 when Albert Einstein developed his theory of the interconvertibility of matter & energy with the famed equation $E=MC_2$, scientists had believed firmly (one might say knew) that matter and energy were two distinct entities. Thankfully, Einstein did not succumb to the pressure of the overwhelming consensus at the time.

Because of our tendency to see through the eyes of prior experience and conditioning, it is clear that observation alone will not fill the bill. We must have a disciplined method of testing our observations, and to ensure objectivity there must be a non-partial review of the test methods and results. Now, we can't all be scientists, so we are left with relying on the word of scientists for truth in the physical world. But here is where we need to be very careful. Scientists are human too, and subject to inadvertent error, and regrettably, occasionally, to temptation to intentionally mislead for purposes of personal gain or power.

While the first step in the scientific method is observation, Karl Popper (1902-94) taught that there were serious limitations to the process of logical induction (drawing general conclusions from what are necessarily limited observations). His famous example of black swans makes the point. For thousands of years Europeans had observed millions of white swans. By induction they might have reasonably proposed a theory that all swans were white. However, that theory was proved false when black swans were discovered in Australasia. The point is that no matter the number of observations that confirm a theory, there always remains the possibility that it could be refuted by a future observation. Induction cannot yield certainty.

Moreover, Popper felt that it was naïve to think that we could objectively observe the world. He contended that all observation is from one's point of view, and all observation is actually colored by our perception. We see the world through the lens of theories we already hold.

Popper proposed a scientific method based on falsification, which by the way is accepted by any true scientist today. However much corroborating evidence there is for a theory, it only takes one conflicting observation to falsify it: only one black swan is needed to repudiate the theory that all swans are white. When a theory is shown to be in error and a new theory is introduced which better explains the phenomena, science advances. For Popper, the scientist should attempt to disprove his/her theory rather than attempt to continually prove it. Popper believed that science can help us progressively approach the truth; but we can never be certain that we have the final explanation.

The scientific method is the technique by which scientists go about developing new insights into the physical and biological world (universe) we live in. There are slight variances based on the particular discipline under study, but it has these features in common:

1. Observation

2. Drawing tentative conclusions from those observations
3. Experimentation or testing those conclusions by further observation
4. Refining those conclusions objectively based on the results of the tests
5. Independent peer review to ensure that there was no personal bias in interpreting the data

The Theorists had done fine with the first two steps, but it was step three, where some began to go awry. The point of step three is not to seek confirmation, but paradoxically, contradiction. Einstein had said *"…all of the experimentation in the world cannot prove me right, but only one experiment could prove me wrong."* It is this sincere pursuit of the contradiction that keeps science honest and leads to a closer understanding of the truth.

Just as there were in ancient times, when science conflicted with prevailing thought, such as the earth revolving around the sun rather than vice versa; or that our planet is a globe rather than a flat plane, there are always those that in spite of overwhelming evidence, will continue to hold to their beliefs. While a mild skepticism with a new theory may be commendable; refusing to even consider an opposing position can get one into trouble, as Mark Twain so pithily put it.

As an example of fanatical devotion, see the following from a dedicated theorist:[1] *"Even if the U.N. came out tomorrow and said 'ok, everyone, we got you good - global warming is a hoax' - I really wouldn't stop believing it unless many respected authors of various studies also admitted it's a hoax, if the Mauna Loa Observatory boffins said their atmospheric carbon dioxide data was a hoax and if my 'spidey sense' said it was a hoax."*

The 'spidey sense' is an important element of all this to me.

In fact, a whole stack of organizations would need to come forward and say their data was a hoax before it would even put a dent in my convictions. I feel the planet dying; I feel the changes - have done since I was a child. I see the changes in plants and animals; I see the difference in the sunrises and sunsets. I feel the winds of major change; a whisper at first and now rapidly building into a storm. Global warming and climate change are only elements of it - we are witnessing a convergence of environmental crises - critical mass, toxic overload; whatever you like to call it."

I am not sure of the exact definition of "spidey sense", but I suppose it works just as well as the more scientific sounding "anomaly" to explain away certain inconvenient challenges to The Theory.

CHAPTER 2
THE THEORY

Now that we are versed in the scientific method, we will take a look at The Theory from the scientific point of view, since after all, it has been presented as scientific truth. The Theory in essence states that carbon dioxide (CO_2) emissions, primarily from the burning of fossil fuels, have caused global temperatures to rise; and that in recent years the increase of global temperatures has accelerated. Moreover, if fossil fuels continue to be used, our planet will soon reach a point (tipping point), at which a global catastrophe is unavoidable. The only prescription to avert this looming disaster is for immediate draconian reductions in the use of fossil fuels.

Supporting this theory are measurements of atmospheric CO_2 levels and global temperatures, as well as proxies for those parameters for historic periods before actual measurements were taken. Further, there have been large numbers of observations of retreating ice caps and glaciers, as well as photographs of the same areas taken over different time periods showing stark differences in snow and ice coverage. Moreover, CO_2 is a known greenhouse gas and is the main gaseous product issuing from the burning of fossil fuels. Lastly, there is supposedly an unmistakable correlation between atmospheric CO2 levels and average global temperatures, tied in to the fact that the increase of CO_2 began concurrently with the start of the industrial revolution when massive quantities of coal began to be used.

At first glance, pretty convincing stuff, at least for non-scientists. The Theory would even appear to follow the general guide lines of the scientific method:

1. Observation (ice caps and glaciers melting.)

2. Tentative hypothesis (Global warming is occurring; and is being caused by the emission and atmospheric retention of large quantities of CO_2.)

3. Testing the hypothesis by further observation. (Measuring global temperatures and atmospheric CO_2 levels, coupled with supposed historic evidence)

4 . Refining of the hypothesis (There would appear to be no refining; but we will let that go; since the theorists apparently feel that the results of the tests confirmed their initial hypothesis.)

5. Independent peer review (Gore and others have pointed out that 2500 scientists have reviewed and concurred with the IPCC statement that anthropogenic global warming is a fact. They have gone on to say that AGW, i.e., The Theory is settled science.

Before we examine The Theory in depth, let's clear up a big misconception. It may shock some theorists and dissenters who are not scientists to hear me say that the anthropogenic warming effect is real. But, and this is a big but, when I say the effect is real, I am not saying anthropogenic global warming is real. Let me explain. As the world's population grows and as more roads and buildings are constructed, there is a very real warming effect. This is anthropogenic, but has nothing to do with CO_2 emissions. There also is no argument that increasing atmospheric CO_2 levels have a warming effect as well. There is a huge argument, however, that the latter effect is greater than that of natural forces. More about that in chapter 4.

But let's now take a deeper look at The Theory, examining each facet through the lens z of the scientific method:

1. Observation- Karl Popper's admonition against relying too heavily on logical induction is key here. Recalling his black swan example, if there be a single contra- indication (and there are many, as you will see in coming chapters), the original hypothesis must be withdrawn and revised to conform with the new evidence.

2. Tentative hypothesis- Where is the falsification aspect? In other words, how would one begin to challenge the assertion? According to The Theory, global warming is occurring as a result of CO_2 buildup in the atmosphere. How would we go about proving it false? How about if we show that average global temperatures have declined in certain periods even while CO_2 levels continue to climb. No, say the theorists. Those periods are anomalies; and we must only consider the long term trends. If we ask what constitutes a long term trend, we do not get a definitive answer.

Nonetheless, the graphs with the apparent upward trend lines that many of us have seen are certainly impressive. We will be taking a hard look at how the numbers that the graphs are based on were determined, as well as the numbers themselves. You can then decide how credible the trends are.

3. Testing the hypothesis by further observation. We will not challenge the CO2 measurements, but we do take strong exception to much of the temperature measurement as you will see in Chapter 3. Likewise, there are problems with the reported historical record revealed in Chapter 5.

4. Refining of the hypothesis. We are conceding a "gimme" as stated above.

5. Independent peer review. Refer back to Chapter 1. Note the qualifiers in this step: "independent" and "no personal bias". Peer also implies that the individual's field of expertise is in the same field being reviewed. We will scrutinize these statements in later chapters. Suffice it to say that there are some big credibility gaps.

Chapter 3
Taking Earth's Temperature

One of the twin premises of THE THEORY is that the earth is heating up and doing so at an unprecedented and accelerating pace since the onset of the industrial revolution in the mid-1800's. The pillar (and a shaky one at that, as you will see in subsequent chapters) that supports this premise is shown by the graph below, which most will recognize as Penn State Professor Michael Mann's famous or infamous (depending on your point of view) hockey stick curve. The moniker, "hockey stick", derives from its similarity in appearance to a hockey stick with a long straight handle and a much shorter upward turning blade.

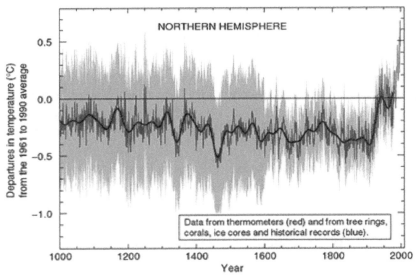

What it purportedly shows is that for at least 850 years prior to the advent of the industrial revolution in the mid- nineteenth century, global temperatures remained essentially unchanged, with only the most minor of fluctuations, but then suddenly turned up at an accelerating slope correlated to an increase in atmospheric carbon dioxide (CO_2) level.

We will deal with some interesting challenges to the curve's validity in a later chapter, but for now we are concerned about the measurement of our planet's temperature; therefore, we will focus on the first part of the statement, "the earth is heating up."

I have a question for you. What was the average global temperature in 1850 when all this began? I don't expect you to know it without checking since I don't know either. But it should be easy enough to find that fact on the internet, shouldn't it?

What? You say you can't find it. Let's make it simpler. How about the average global temperature for 1900 or 1950 or 2000? Don't bother looking. You will not find it.

We have heard that the average temperature of earth is about 59^0 F, but that is only an estimate. I guarantee that you will not find a definitive answer to an average global temperature for any year. Why? Simply because we lack the ability to measure the average temperature of our planet, considering that those temperatures are constantly changing and that we cannot monitor every point of the earth continuously. As proof of my statement, consider that from time to time the NOAA (National Oceanic and Atmospheric Administration) revises their supposed irrefutable temperature data. 1998 had been declared as the hottest year in the US, but in 2007, it was changed to 1934.

Further complicating the matter are the wide variations in temperature around the world at any given point in time. One then might reasonably ask how we can determine that our planet is any warmer or cooler than the day before, let alone, years or decades earlier. Well, approaching it from a scientific point of view, we would want to set up temperature measuring stations at a number of selected points around the world. The number of points, being a statistical issue, we will leave for discussion in chapter 12. Since measurement of temperature is our concern right now, let's examine the selection process.

Intuition tells us that we would not want to concentrate the measuring too heavily in one area, for fear of distorting the results. The converse is obvious as well; we should strive to have the points distributed as evenly as possible over the surface of the globe. Therein lies the first problem. Some areas are just not accessible because of topography. Ideally, we would place all the measuring stations at the same height, at or slightly above sea level, so as not to distort the data by natural variations in temperature with altitude. Other areas may be off limits because of war, threats of violence or other situations, such as radioactive hot spots or live volcanoes. We are thus relegated to a relatively small portion of the earth's surface.

Stations with Temperature

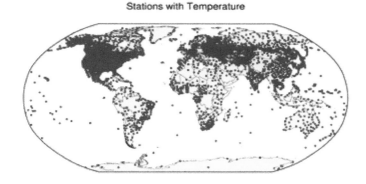

Each point signifies a recording station, and it is clear that the distribution leaves a bit to be desired. Distribution is not the only problem, however, since the number of recording stations has changed significantly over the last century as you will see below. This change in number and location of recording stations presents a second problem; but not to fear, the theorists have answers to both of the problems. The altitude variations and uneven distribution distortions are "corrected for" by dividing the globe into cells and averaging the temperatures of stations within the cell so that cells with more stations are not more heavily weighted than those with fewer stations. NOAA and Hadley Climate Research Unit (Had-Cru) divide the globe into 5184 5'x5' cells, whereas GISS, apparently recognizing the 5x5 approach creates cells of differing size (the closer to the poles, the smaller the cell), divides the globe into 8000 equal area cells. All three agencies use the same temperature recording stations; so the GISS approach results in more empty cells. I have a suggestion that would resolve both of those issues and be a lot simpler to boot: Just use four recording stations: one at each pole and two at the equator spaced at 180⁰ longitude apart.

As for the time and location distortions, they are likewise "corrected for" by the use of temperature anomalies (deviations from an established base line) rather than absolute temperatures. We will examine the interesting concept of anomalies in a later chapter.

So, apparently, we need not concern ourselves with either problem? I hate to be a nit-picker, but I see a slight problem with both of the above corrections. The cellular approach would appear to be a reasonable approach to eliminate the distribution problem except for one thing. There are a large number of cells with no measuring stations. In fact, there are far more empty cells than there are that contain at least one station. Even in the years having the greatest number of cells with stations, only 16+% of the cells contained stations, in 1998 it was about 10%. See graphs below. [1]

Now, you may be curious as to how I came up with the percentages cited above, since the graph in figure 2(b) looks as though the percentages with filled cells should be 85% and 50%, respectively. However, if you take a closer look, you will see that the graph shows only 1000 cells. One has to wonder if the individual who presented the graph in such fashion just forgot the other 4184 cells, or perish the thought, possibly had some hidden agenda.

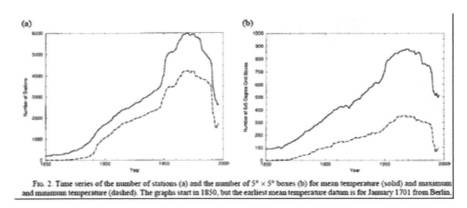

Fıg. 2. Time series of the number of stations (a) and the number of 5° × 5° boxes (b) for mean temperature (solid) and maximum and minimum temperature (dashed). The graphs start in 1850, but the earliest mean temperature datum is for January 1701 from Berlin.

In light of the following, I find the exchange on Duke Professor Bill Chameides's website "The Green Grok"[2] particularly interesting. A blogger asks Dr. Chameides how he can make the statement that 1998 is the hottest year on record, when NASA itself has acknowledged 1934 to be the hottest. The learned professor admonishes the blogger to check his facts, i.e. 1934 was US's hottest year; but 1998 was the hottest globally. Say what?! I thought uneven distribution was compensated for by the use of cells. After all, there is no distribution problem in the US. The rest of the world accounts for all the empty cells. So how global is "globally"? I wonder with what certainty he believes 1998 to be the hottest year. I gather that our esteemed professor does not see the irony in his statement or just prefers to ignore the contradiction.

Moreover, as mentioned above, the stations have not remained the same, either in location or number. Take a look at the changes from 1900 to 2000.[3]

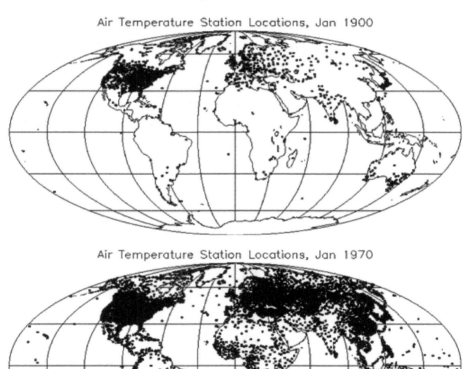

Air Temperature Station Locations, Jan 1900

Air Temperature Station Locations, Jan 1970

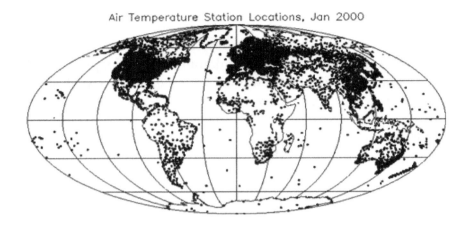

Air Temperature Station Locations, Jan 2000

Comparison of Available Global Historical Climate Network (GHCN) Temperature Stations Over Time

The following figure shows the number of stations for selected years, showing the number of stations in the United States (blue) and in the rest of the world (ROW – green). The percents indicate the percent of the total number of stations that are in the U.S.[4]

US temperature stations vs. rest of world stations over time

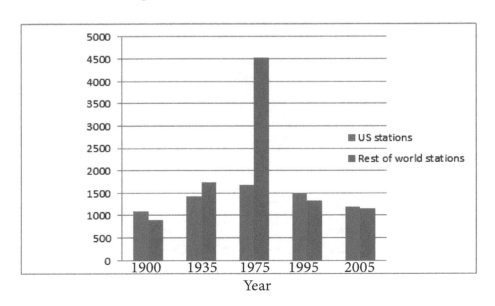

Comparison of Number of GHCN Temperature Stations in the U.S. versus Rest of the World

I hate to sound like a conspiracist, but one must ask why the numbers of stations are being reduced. I understand that certain locations may become untenable because of war or other circumstances, but a reduction of what appears to be about 60% from 1975 to 2005? Isn't the more data, the better? Again, is there perhaps a hidden motive? As in selecting the right (warming) data?

Apparently there was not enough warming in the AZ 5^0 cell, which was changed from 1991 to include Phoenix in 1994.[5] Also in 1994, West Palm Beach was added to the FL 50 cell in preference to the rural areas of Everglades and Arcadia, both of which have exhibited cooling trends.[6]

So we are to understand that temperature anomalies correct for the changes. Maybe I am missing something, but I fail to see how one can establish a base line where there was no prior history.

Take one of the many points in east South Africa in 1976, for example. As you can see in the 1900 map, the stations in that area are quite sparse. How do you assign a baseline temperature for a nonexistent station? I am sure the theorists have an answer to that question as well; just extrapolate. Am I the only one that might question the accuracy of such extrapolation, when there were only six South African stations in 1900 for an area of over 470,000 square miles?

OK, enough of that; I don't want to beat a dead horse. But there is another and equally serious problem with surface temperature measurement and that is what are referred to as urban heat islands. It is a well-known and documented fact that cities are warmer than nearby rural areas, and the larger the city, the warmer. Buildings and pavement, industry and people all contribute to warming; so that if an area has changed from rural to urban, there will be a consequent upward temperature trend as compared to surrounding undeveloped areas.

With that in mind, Anthony Watts of the "Watts Up With That" website fame decided to visit some nearby temperature monitoring stations, and what he found shocked him. Based on his initial findings, he started another website documenting and photographing the National Climatic Data Center's (NCDC) temperature measuring stations. As of 2011, 70% of the surveyed sites had such flagrant violations of NCDC's own siting protocol that they were in the 2^0C to $>5^0$C error range.[7] Recalling that NOAA states that global temperatures over the last century have gone up 0.6^0 C, I have to ask, can they be serious?

Below is just one example. As you can see, the temperature spiked sharply upward in 1981, which coincided with the construction of the nearby condominiums, and more likely than not, when they started burning refuse there. [8]

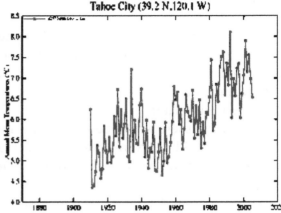

Since the data are transmitted by satellite directly to the NCDC in Asheville NC, the personnel there may well have been unaware of the erroneous data coming from such poor monitoring stations; but NCDC's protocol requires annual site inspections and maintenance.[9] Perhaps since the trend went in the expected direction, no one felt that there was any need for inspections; but one must wonder what the response would have been if the curve had spiked downward rather than upward. Nevertheless, I would guess that since this information appeared on Watts' website, the most egregious violations have been corrected; but again there is the question of how the record is being handled. Let's wait and see if the recent US temperature record is revised downward.

As of now (June 29, 2011), Watts and his group of volunteers have documented with photographs 82.5 % of the stations in the US. Below is a chart showing the NASA station ratings according to NASA's own protocol with over 70% of the stations surveyed having an uncertainty of 2^0 C or greater.[10]

With all the aforementioned problems, guesswork, and just poor science associated with surface temperature measurement, one might reasonably conclude that there is no way to reliably measure earth's temperature. However, there is hope. Since 1979, UAH (University of Alabama at Huntsville) in association with NASA has been taking temperature via satellite. There are limitations for sure such as not being able to cover the whole earth at the same time. Nonetheless, it eliminates most of the problems associated with ground measurement.

The satellite data through 2009 shows little if any warming globally, and more importantly in the lower tropical troposphere; since according to the IPCC models, this is the part of the atmosphere where any warming would be felt the most.[11] The conclusion is clear: The models, which on average show twice the observed temperature trend; and upon which much of The Theory is based are severely flawed.

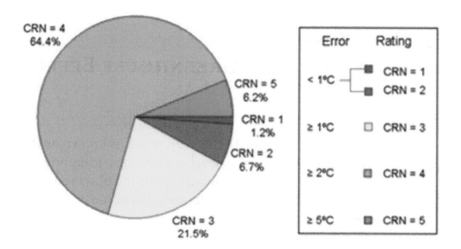

Moreover, the surface temperature record itself betrays another inconvenient fact. Since 1850 the southern hemisphere has shown an average of about one half the temperature increase as the northern half.[12] Might the urban heat island effect have something to do with the difference in hemispheric warming, given the fact that there are more and larger cities in the north than in the south? [13]

And finally, in a candid admission by the NOAA, " One of the principal conclusions of the 1997 Conference on the World Climate Research Programme was that the global capacity to observe the Earth's climate system is inadequate.… In spite of the United States being a leader in climate research, we do not have, in fact, an observing network capable of ensuring long-term climate records free of time-dependent biases. Even small biases can alter the interpretation of decadal climate variability and change." [14]

CHAPTER 4
GREENHOUSE GASSES AND THE GREENHOUSE EFFECT

Now we come to the other of the twin premises of The Theory, which is known as the greenhouse effect. The premise here is that the increasing level of atmospheric carbon dioxide (CO_2) since the beginning of the industrial age is responsible for the observed global warming discussed in the previous chapter. The pillar that supports this premise, while sturdier than the first, has its cracks as well.

The argument is that CO_2 is a known greenhouse gas and is the main combustion product of fossil fuels. When the industrial revolution began in the mid- nineteenth century, coal began to be consumed in tremendously increasing quantities, followed and added to by petroleum when the automobile was invented. Atmospheric levels of CO2 have been determined to have been at stable levels for millennia until a sharp rise began in tandem with the increased burning of fossil fuels and the increase in global temperatures.

It may shock some non-scientist theorists to read this, but dissenting scientists do not dispute much of the above. CO_2 is indeed a greenhouse gas and is the main product formed from the combustion of fossil fuels in an oxygen rich environment, which air is. We agree also that atmospheric CO_2 levels have just about doubled from what they were in 1850.

We disagree with the statement that CO_2 levels have been stable for thousands of years and will discuss this further in the next chapter on proxies. Moreover, we do not believe that CO_2 is the main actor in whatever warming we may have experienced in the twentieth century and recent years. Indeed, most of us are not convinced that atmospheric CO_2 levels have had any significant effect on global temperatures.

First, let us address the greenhouse effect. Actually, the use of greenhouse as an analogy is misleading since most of a greenhouse's heat trapping ability derives from its glass walls preventing convection. If a small window at the top of a greenhouse is opened, there will be a significant drop in temperature due to the warm light air rising through the opening and being displaced by the denser cold air. In the atmosphere, with convection free to operate, the global warming effect is countered by convection. Therefore, only a portion of the heat buildup on earth is due to what is commonly known as the "greenhouse effect."

Approximately half the energy of the sun's radiation is visible light, which passes through the atmosphere undisturbed. The other half consists of a small portion of ultra violet radiation (UV), most of which is blocked by the ozone layer, and the remainder infra- red radiation (IR), much of which is absorbed and re-radiated in all directions by the greenhouse gases. When the visible light reaches the earth, an estimated thirty percent is reflected back as visible light, again passing unimpeded through the atmosphere and into space. The light that is not reflected is absorbed and is the primary source of heat for the earth. As the earth's surface heats up, it in turn emits IR into the atmosphere, and just as occurred with the incoming IR, much is absorbed and re-radiated by the greenhouse gases. It has been calculated that earth's temperature is 59' F warmer than it would be, absent greenhouse gasses. This fact is undisputed.

There are two points to be made, however. First, remember the window in the greenhouse. Convection may be thought of as the earth's thermostat. It is technically incorrect to liken the atmospheric greenhouse effect to a blanket, unless we qualify the term and refer to a radiative blanket. As the atmospheric level of greenhouse gasses increases, increasing amounts of IR energy are trapped, but because one can only approach 100%, each incremental increase of greenhouse gas becomes less and less efficient. On the other hand, as the lower air becomes warmer it also becomes lighter and moves more strongly up and out the "greenhouse window".

Second, CO_2 is only one of several greenhouse gasses. Several other natural and man-made gases act similarly. Methane, carbon monoxide, CFC's, sulfur dioxide, nitrous oxide, water vapor, and ozone are all greenhouse gases. If I were to ask which of these are the most important, a lay person could be forgiven for answering CO_2. In fact, the gas having the greatest greenhouse effect in our atmosphere is water vapor; and the comparison is not even close. This fact too is undisputed, even among the scientists who are Theorists; but they obfuscate the subject by concurrently talking about forcings and feedbacks; and pointing out that water vapor is different in that it condenses.

Moreover, water vapor concentrations are widely variable around the globe and even vary widely in a given area on a daily basis. All of this is true; but none of it changes the basic fact that CO_2's greenhouse effect is miniscule as compared to that of water vapor.

CHAPTER 5
PROXIES

So, the theorists tell us that for at least 2000 years prior to the advent of the industrial revolution in the mid- nineteenth century, global temperatures remained essentially unchanged, and atmospheric levels of CO_2 have been determined to have been at stable levels for a similar time until the onset of the industrial age. We dissenters dispute both propositions; so how do scientists on either end of the argument substantiate their claims in view of the fact that the thermometer was not invented until 1593, and the earliest continuing temperature data is from Central England in 1659. Moreover, chemical analysis of CO_2 only began in 1812, and instrumental records only date back to 1958. Compounding the ambiguity is the fact that most of the data from chemical analysis shows that atmospheric CO_2 levels were anything but highly stable, varying markedly over the years from 1812 to 1958. The data showed maxima in 1825, 1859, and 1942, and the 1942 readings showed levels in excess of 400 ppm.[1]

"Not to worry", say the theorists. They have deemed the preponderance of that data flawed and inaccurate. Interestingly, the relatively few points that were accepted fit neatly into the gently upward sloping curve that we have all seen. So what was the problem with the bulk of the data? It didn't conform to the curve; therefore, it was determined to be erroneous. That is very interesting science. I maybe could have used that technique in quantitative analysis when my three readings were not consistent. When two were close, I could have assumed the outlier was erroneous and merely discarded it. If the three were spread widely, I could have just averaged them, considering the high and low values to be erroneous. The only problem is that my professor would likely have rewarded my ingenuity with an F.

It should be pointed out much of the excluded data came from Nobel Laureates. Come to think about it, considering the credentials of certain recent Nobel Prize recipients, perhaps that is reason enough to eliminate them from consideration.

The question then is how can scientists make any definitive statements about climactic conditions prior to the time of direct measurement. The answer is that there is evidence from natural archives such as ice cores, tree rings, bore holes, corals and plant stomata; all of which are referred to as climate proxies. The deposition or growth rate of the proxies' material is dictated by the climate conditions when the deposition or growth began.

Traces of chemical isotopes recovered from proxies may also offer clues about changing climate conditions and fluctuations in the composition of the earth's atmosphere. The presence of animal or plant fossils at a latitude or altitude where that species does not currently exist may indicate changes in climactic conditions as well. Since direct observation is not involved, however, one might compare climate proxies to circumstantial evidence in a trial. Just as in a trial, where it is more difficult to convict with circumstantial evidence, corroborating evidence is crucial.

In view of the ambiguities, let us examine each of the above proxies to determine their credibility:

Ice core samples- When it snows, air is dissolved in the falling snow, providing a metaphorical snapshot of the air's composition. As layers of snow and ice build up, there is then a record of changes in that composition. But how reliable is that record? Their use as a proxy for historic atmospheric CO_2 levels is based on the premise of a closed system, i.e., that at any period of historic time what is captured in that layer of ice is there until a scientist examines it. That premise is severely flawed, however, because it does not take into account that liquid water may be entrained in very cold ice under the high pressures of glacial ice.[2] The water then acts as a conduit through the layers; and since CO_2 is tremendously soluble in cold water, the gas is evenly spread through the layers. This explains the stable CO2 levels found over the last 1000 years, in spite of significant fluctuations in average global temperatures over that period. Dr. Zbigniew Jaworowski, a world renowned multi-disciplined scientist with particular expertise in glaciology pointed that out as well as some other problems in ice core analysis in testimony before Congress in 2004.[3]

Under great pressure and cold temperatures CO_2, methane, and other gasses combine with water and ice to form ice like compositions called clathrates.[4] When an ice core is drilled, the pressure is released and the clathrates explode like miniature grenades, further contaminating the layers and distorting the data. Furthermore, the aging of ice cores is questionable, the further back in time due to the extreme pressures exerted on deep ice.

Isotopic compositions of water, involving "heavy oxygen" (^{18}O) and "heavy hydrogen" (^{2}H), otherwise known as deuterium, are used to help determine temperature changes. Water vapor made up of the heavy isotopes condenses more rapidly and then freezes to yield a higher percentage of "heavy" ice under relatively warmer conditions.[5] The different compositions in each layer are then presumed to mirror temperature changes over the ages. This process again ignores the presence of liquid water in the ice, which theoretically should transport the heavier isotopes downward; leading to the conclusion that more recent ice of several hundred years is from a cooler period and the follow up conclusion that present day temperatures are relatively warmer.

Normally, ice cores from Greenland are analyzed for ^{18}O, while those from Antarctica are analyzed for ^{2}H. Those cores analyzed for both show a lack of agreement.[6] This is another example of shoddy science. Every scientist knows that when you repeat an experiment, using the same technique, and obtain different results, there is something wrong with the experiment or with the data, or both. As a matter of fact, any first year science student (who passes the course) learns this.

Proxies

Tree rings- As deciduous trees grow, each year they develop a ring; so it is possible to determine the age of a tree by counting the rings. The closer the rings, the less growth there was for that season; and conversely, wide spaces between rings indicate strong growth. Tree rings then serve as a proxy for growing conditions, i.e., more days of sunshine without frost, higher CO_2 levels, and adequate moisture all favor faster growth. But how do we know which set of circumstances took place? Also might other factors such as fungus or insects inhibit growth in otherwise favorable conditions? Moreover, in the tropics where seasons don't change all that much, it is difficult to impossible to get an accurate ring count. This means that tree selection is pretty much confined to the temperate zones, where the seasons differ more markedly. Furthermore, since less than 30% of the earth is dry land, tree selection starts out by excluding more than 70% of the globe.

This brings to mind the story of the drunk looking for his wallet under the street light. When the cop asks him if he recalls specifically where he might have dropped it, the drunk replies, *"Why, way over there."* The cop then asks why if he dropped it *"over there"*, he is looking *"over here under the streetlight."* The drunk replies, *"Simple, there's a lot more light over here."*

Bore Holes- The underlying science behind the use of bore hole temperatures as temperature proxies is the slow heat transfer from the surface, resulting in temperature differentials at incremental depths, supposedly corresponding to changes in surface temperature over time. The science is fraught with assumptions, uncertainties, and problems, the first of which is that the earth's core is very hot, requiring complicated mathematics to compensate for the rising heat from the core. This mathematical process yields a "solution" with a "non-unique" series of surface temperature values. It is "non-unique" because multiple possible surface temperature profiles can result in the same borehole temperature profile. Moreover, due to physical limitations, the reconstructions are inevitably distorted and become more so with time.

"When reconstructing temperatures around 1,500 AD, boreholes have a temporal resolution of a few centuries. At the start of the 20th Century, their resolution is a few decades; hence they do not provide a useful check on the instrumental analysis record."[7]

Another major problem with the bore hole technique is the assumption of a homogeneous medium over the various depths of the hole, which geologists will tell you is rarely the case. Since different substances differ in heat conductance, readings can be misleading. Subsurface water can introduce even greater distortion. Ice core bore holes where the site remains continuously frozen do not have the homogeneity problem, but still encounter the required compensation for earth core heat. And we are back to the problem of sample selection over a small portion of the earth.

One more interesting tidbit before we leave boreholes. Borehole data broadly confirm the prevailing belief that average global temperatures have increased 0.70 C over the last century; so my guess is that theorists feel the science is "robust", to use one of their favorite words. I wonder how robust they feel the following statement from "Surface Temperature Reconstructions for the Last 2000 Years" is: (The bold is my insertion.)

*"For central Greenland (Cuffey et al. 1995, Cuffey and Clow 1997, Dahl-Jensen et al. 1998), results show a warming over the last 150 years of approximately 1°C ± 0.2°C preceded by a few centuries of cool conditions. **Preceding this was a warm period centered around A.D. 1000, which was warmer than the late 20th century by approximately 1°C.** An analysis for south-central Greenland (Dahl-Jensen et al. 1998) shows the same pattern of warming and cooling but with larger magnitude changes. Uncertainties on these earlier numbers are a few tenths of a degree Celsius for averages over a few centuries."* [8]

Corals- Tiny colony dwelling marine organisms mostly found in tropical and sub-tropical oceans and seas around the world secrete a limestone or calcium carbonate exoskeleton, referred to as coral. Rings or bands found in certain varieties of coral may harbor clues to paleo-climactic conditions, similar to tree rings. Cold temperatures are known to slow the metabolism of the organisms, leading to narrower bands and higher density limestone. The converse is true under more favorable conditions. That would indicate a type of paleo-thermometer, except that the situation is far from as simple as a direct thermal relationship. Just as with tree rings, there are other conditions that affect the rate of calcium carbonate deposition. It so happens that calcium carbonate, although considered insoluble, is slightly soluble in water; and as with most salts, the solubility increases as the temperature rises. The increase is only slight, but it should not be ignored since it will impact limestone deposition rates. Moreover, solubility is affected by the presence of other substances. Most laymen think of the ocean as just "salt water" and in a sense this is correct, although certainly simplified. In reality, the oceans and seas are a complex chemical "soup", consisting of at least seventy-five per-cent of the elements. When I say "at least seventy-five per-cent" it means that is the percentage that has been detected. As more sensitive analytic procedures are developed, it may be assumed that more of the elements will be found; since, after all, water is known as the "universal solvent". Nevertheless, 99.9 percent of this chemical soup is a solution of eleven ions[9], which interact to form a great many more substances.

The secretion of the limestone skeleton is a biochemical process heavily dependent on the relative solubility/insolubility of calcium carbonate. And that solubility index is determined primarily by four factors: calcium ion concentration, bicarbonate ion concentration, pH, and temperature. But there are other factors that enter into the game, sulfate and phosphate ions as examples. Since calcium sulfate and calcium phosphate are more insoluble than calcium carbonate, increases in sulfate and phosphate ion concentration are accompanied by a reduction in calcium ion concentration. As a consequence, calcium carbonate becomes more soluble, which in turn slows the rate of growth of the exoskeleton. A competing factor impacting calcium carbonate solubility in the oceans is run-off from terrestrial limestone, which has been dissolved by acid rain. When this additional calcium bicarbonate reaches the ocean, calcium sulfate and calcium phosphate are precipitated, raising the pH and concurrently reducing calcium carbonate solubility.

The average ocean chemical composition is remarkably constant; but there are significant regional variations in salinity and large spot differences in bicarbonate and sulfate concentrations as well as pH due to volcanic eruptions. In a recent work by oceanographers, Hillier and Watts from 2007, the two had surveyed 201,055 submarine volcanoes in a small area of the Pacific Ocean equivalent in size to the state of Maryland, and projecting that to the vast area of the oceans estimated that there were in excess of 3,000,000 submarine volcanoes world- wide.[10] Timothy Casey, a geology consultant, conservatively estimates four percent of those to be active, based on observations by Rodey Batiza, in his paper, *"Abundances, distribution and sizes of volcanoes in the Pacific Ocean and implications for the origin of non-hotspot volcanoes"*.

Affecting the paleo-thermometer of coral (which supposedly is a recorder of surface air temperatures) as well is the tremendous heat released by submarine volcanoes and hydrothermal vents. Temperatures of water emerging from hydrothermal vents ranging from 600C to 4640C have been recorded.[11]

Before we leave corals, a couple of thoughts have occurred to me. 1. Biologists know that all life evolves and adapts to different environments. How do we know that corals have not adapted as well so that changes in temperature have muted changes in metabolic rate? And 2. With the wide diversity of coral organisms, can we assume that they all have the same metabolic rate? With these issues and all of the other variables affecting coral growth, one must question the reliability in their use as bio-thermometers.

Stomata are pores in the leaf and stem epidermis of plants, used for gas exchange. .Carbon dioxide and oxygen enter the plant through these pores where it is used in photosynthesis and respiration, respectively. Their use as paleo-proxies is based on an inverse relationship between historic CO_2 levels and size of pores. As with the other proxies, there are inconsistencies, but there do not appear to be any more incongruities with this proxy than with the others. That this evidence of past periods of significantly higher CO_2 levels than present[12] is pretty much ignored or discounted by the theorists; I guess we are to assume is mere coincidence.

We have only covered the most popular of the proxies; there are others, and as I write, there are probably more being added. In each of the proxies, we have seen that there are a host of variables, known and unknown, which affect the presumed paleo-evidence.

Proxies may be likened to a piece of evidence from a contaminated murder scene which has been discovered a considerable time after the crime took place. The confusion may be compounded by various elements, including contradictory evidence, multiple suspects, conflicting testimony, and inconclusive cause of death.

Think back to unsolved murder cases like that of Jon Benet Ramsey[13] and Natalee Holloway[14]. In these cases as in most unsolved crimes there was no eye witness unless occasionally, there are conflicting accounts. In Ramsey's case there were initially two prime suspects (her brother and her father) and possibly others from the Ramseys' wide circle of friends. Later, DNA evidence indicated that there may have been as yet an unidentified intruder. Also, the murder scene had been disturbed and contaminated. In Holloway's case all circumstantial evidence pointed to Joran van der Sloot, but her body was never found.

Neither case ever went to trial, but let's imagine that the Holloway case had been tried and you were a member of the jury. Could you be certain beyond a reasonable doubt that Natalee had even died? Isn't there the possibility that she had been sold into the white slave trade? And even if van der Sloot had killed Natalee, could you be sure the act was intentional?

One might think of theorists as prosecutors trying to convince us that proxies prove that certain events from long ago occurred. But how can we or they be certain when there is conflicting evidence that indicates other possibilities?

Chapter 6
Forcings and Feedbacks

Forcings are the direct effects on aspects of climate, such as solar energy, volcano eruptions, and the greenhouse effect discussed in chapter four. Feedbacks are the indirect effects occurring as a result of the forcings. These feedbacks may be positive or reinforcing such as melting ice caps, which reduces albedo (defined below), or rising ocean temperatures, which lowers CO2 solubility and in turn increases atmospheric CO_2 levels leading to more warming. Or negative, as in cloud formation, which increases albedo. The theorists appear to have a predisposition toward positive feedbacks as they tend to ignore or downplay several of the negative variety. An example is the emphasis on evaporation caused by increased temperatures, which in turn causes more warming, since water vapor is a greenhouse gas. Clouds also are a result of evaporation, but they don't seem to get as much press. Could it possibly be because clouds generally contribute to cooling, by shading and reflection of sunlight?

As stated, greenhouse gases are forcing agents because they contribute to heating the earth; but they also act as feedback mechanisms. Since gases are less soluble in warm water, the oceans release dissolved CO_2 as temperatures rise. The additional CO_2 results in higher temperatures causing more CO_2 to be released, etc.

Likewise, methane is prevalent in permafrost; and as temperatures rise, the permafrost melts and releases methane. The added methane in the atmosphere similarly contributes to higher temperatures. From these two examples, it is easy to see why many are concerned. After all, it would appear that once that feedback is set in motion, there would be no stopping it as temperatures get higher and higher till they reach and surpass catastrophic levels. Thankfully, the earth is more resilient than that.

First, one can look at an analogy. Increasing greenhouse gases may be compared to piling on the blankets when you are cold. The first blanket warms you by maybe holding in 60% of your body heat. As you add successive blankets, you will get closer to 100%, but you will never raise the temperature above your body temperature (unless, of course you use an electric blanket). So, there is a positive feedback, but it is self-limiting.

Theorists will argue that my analogy is invalid, because I have ignored the sun, which could be likened to the electric blanket. OK, I will concede the point, but I think in as much as they are bringing up omissions it is only fair for me to do likewise. Therefore, I should point out that they are ignoring the "open window in the greenhouse", discussed in chapter four, and which in my example could be likened to leaving part of your body uncovered.

The atmosphere currently absorbs about 96% of the reflected infrared radiation,[1] and remember, water vapor is a far stronger greenhouse gas than CO_2. So, how much of an effect on temperature could further increases in CO_2 have? Moreover, as pointed out, convection plays an important part in moderating the temperature increase.

Albedo is defined as "as the ratio of reflected radiation from the surface to incident radiation upon it."[2] More simply, it is the reflecting power of a surface. The classical climate feedback involving albedo occurs when increasing temperatures causes snow to melt, resulting in lower albedo because water's albedo is much lower than ice. (More energy is therefore absorbed, driving temperatures higher.) It is interesting that the melting process itself absorbs heat, which goes counter to the albedo effect; but this is rarely mentioned.

Another negative feedback is occurring in the Arctic; where as tundra melts, new forests are developing, and in the growth process are consuming large amounts of atmospheric CO_2 through photosynthesis, discussed below. But, not to worry, the theorists have a ready response. *"Forests will expand northward into the current tundra regions. Although forest growth increases carbon dioxide uptake, this beneficial effect will be* **overwhelmed** *(my bold) by the release of large stores of methane and carbon dioxide as tundra regions thaw. The increased absorption of solar radiation by forests, compared to the more reflective tundra they will replace, will also lead to net warming."*[3]

Overwhelmed? That is an overwhelming thought. I wonder how the theorists know this or can prove this, but I guess we are not to question concerned scientists.

Yet, another related negative feedback from Antarctica has only recently been discovered. Open water resulting from melting of ice shelves and glaciers around the Antarctic Peninsula is harboring large quantities of new marine plant growth.[4] No response yet from the theorists, but don't worry; I am sure one will be forthcoming.

Another feedback that appeals to the theorists is evaporation. As temperatures rise, evaporation increases, leading to more water vapor in the atmosphere, which in turn further increases temperature through the greenhouse effect. However, it still rains, no matter how hot it is, since water vapor is less dense than nitrogen and oxygen, and therefore will continue to rise until it encounters cold air, at which point it will condense.

Photosynthesis is one more important feedback, albeit negative, which surprise, surprise, gets little press. It is a known scientific fact that CO_2 is an essential component in the growth of green plants. By the process of photosynthesis, the chlorophyll in green plants allows them to use solar energy to synthesize cellulose and starch from CO_2 and water, thereby removing CO_2 from the atmosphere.[5] Moreover, it is well documented that CO_2 acts like a natural fertilizer and encourages growth, leading to more CO_2 absorption.[6] The theorists acknowledge this effect in the converse by stating that burning and clearing forests contribute to global warming by reducing this effect.

Aerosols are one more negative feedback. Just as greenhouse gases contribute to global warming by retaining heat, atmospheric aerosols operate in the opposite direction by increasing earth's albedo and reflecting solar energy back into space before it even reaches earth.[7] Aerosols are formed when certain chemical compounds react with water. Among the most well- known aerosol creating compounds is sulfur dioxide, which comes from volcanoes as well as industry. Much of the industrial production of aerosols results from the burning of coal. Yes, you read it right. Aerosols (smoke) created by the burning of coal contain a substance that has a depressing effect on temperature. I wonder if we can come up with a name for this effect. How about … anthropogenic global cooling?

Chapter 7
Corollaries of the Theory

Corollary- An accepted truth derived from that which is proven or given.

All good theories need corollaries, I guess. And The Theory is not lacking there. In the previous chapter we covered many of the climate feedbacks. What all the feedbacks have in common are their supposed propensity to move into a runaway situation after a certain point is reached. This point is referred to as the tipping point and is the most important corollary of the Theory.

What it states is that as the earth continues to heat up, the feedbacks associated with global warming, i.e., increased humidity through evaporation, reduced albedo as ice packs melt, etc. will result in an ever increasing rate of heating. At some point, the tipping point is reached when we are into a new climate paradigm with all sorts of undefined catastrophic events occurring, perhaps even the end of life on earth. This makes for great science fiction, but ignores the resilience of nature. For sure, there are certain catastrophic events that could maybe trigger a tipping point such as a giant meteor hitting the earth or nuclear war kicking up such a dust storm that would put us in darkness for years. This could conceivably kill green plants that depend on the sun's energy, and in turn animals that feed on those green plants and so on.

But The Theory envisions a climate change due to a build- up of CO_2 over a period of years rather than an instantaneous event, where plants and animals would have time to adapt. Moreover, The Theory ignores the negative feedback of clouds and faster growing green plants which would thrive on the higher CO_2 levels and in turn consume more CO_2.

Also ignored is the ocean's almost insatiable appetite for CO_2. As CO2 increases, vapor pressure equilibrium increases the ocean's capacity to absorb the gas, understanding that there is a conflicting effect in that as the water gets hotter it decreases the absorptive capacity. But then the additional water from melting ice counters that by providing more absorptive capacity.

A second important corollary is severe weather. Never mind that storms result from temperature and/or pressure differentials; not high temperature per se. When it comes to making a point, the theorists often use the planet Venus as an example. In this case they can allude to the terribly violent weather which occurs on Venus. There is only one problem, however. All of that violent weather occurs high in Venus' atmosphere, which has tremendous vertical temperature differentials. The surface of Venus, which is a stable 800 degrees F, is remarkably calm. *"Any variability in the weather on <the surface of> Venus is noteworthy because the planet has so many features to keep atmospheric conditions the same..."*[1]

So the argument goes back and forth. The theorists maintain that global warming (or global climate change) is responsible for more severe & stronger storm events; but what does the record show?

Among many of the claims by the theorists these days is that global warming may be responsible for the recent record snow storms. I guess global warming is also responsible for the record cold weather. They are the ones that probably also believe that hot water freezes faster than cold water.

The devastating hurricane, Katrina, in 2005, became a rallying cry for the theorists. It, along with cyclones Wilma also in 2005, Larry and Glenda in 2006, was proof that global warming was causing severe weather; and that we could expect even more powerful and destructive hurricanes over the next few years. And in their "Alice in Wonderland" type of circular logic, these severe storms were proof positive of global warming. Since then (as of Nov 2011), there has not been a hurricane or cyclone close to the strength of those cited. The theorists may argue that the reason is that global temperatures have not gone up since then. Really? I thought temperatures were supposedly correlated with atmospheric CO_2 levels, which are certainly higher than five years ago.

Dr. Ryan Maue, a meteorologist at Florida State University, has studied hurricanes and cyclones (the designation for hurricanes in the southern hemisphere, where they rotate counter clockwise instead of the reverse) for years. He states that the way to gauge hurricane strength is to measure maximum wind speed and duration. According to these measurements, there has been no increase in total energy worldwide over the last forty years. Moreover, in the six years since Katrina, global hurricane and cyclone frequency and strength have gone down significantly, and as of August 29, 2011, were at near historic lows.

Dr. Maue, although not alone in his conclusions however, is just dismissed as another of the "few" deniers (dissenters). But then, the theorists switch the discussion to the number of severe storms. So, are we having more severe weather? Not according to an article in the journal Meteorology and Atmospheric Physics. *"Any changes associated with warming of the surface compared to a smaller temperature rise in the lower-troposphere (and resultant changes in atmospheric stability) have not produced detectable impacts on intensification rates of tropical cyclones in the North Atlantic basin."*[2]

Before we leave the subject of the severe weather corollary, it might be of interest to refer to the dates of the ten strongest hurricanes in US history[3], and then consider how closely they correlate to global warming:

1. The Great Labor Day Storm September 2, 1935 Florida minimum pressure 892 millibars.

2. Hurricane Katrina August 2005 Louisiana and Mississippi minimum pressure 904 millibars.

3. Hurricane Camille August 17-22, 1969 Mississippi, SE Louisiana, and Virginia minimum pressure 909 millibars.

4. 4 Hurricane Andrew. August 24 - 28, 1992 Florida and Louisiana minimum pressure 922 millibars.

5. Unnamed Hurricane August 29, 1886 Indianola, Texas minimum pressure of 925 millibars.

6. The Atlantic-Gulf Hurricane September 10 - 14, 1919 Florida, Texas minimum pressure 927 millibars.

7. San Felipe-Okeechobee Hurricane September 16 - 17, 1928 Florida minimum pressure of 929 millibars.

8. Hurricane Donna September 8 - 13, 1960 Florida to New England minimum pressure 930 milibars.

9. Unnamed Storm September 30, 1915 New Orleans, Louisiana minimum pressure 931 millibars..

10. Hurricane Carla September 11, 1961 Texas minimum pressure 931 millibars

Some may question my omission of Hurricane Sandy in the above list of hurricanes, given the devastation it caused along the east coast in October 2012. The answer is simple- for all the havoc it raised, Sandy was not a tremendously strong hurricane. True, it was the second costliest hurricane in US history at $85 billion and it resulted in 285 deaths, but it was only a category three storm with a minimum pressure of 940 millibars and maximum wind speed 115 mph at its most intense, when it first made landfall in Cuba. By the time it hit the northeast coast, it had degraded into a category two storm. What changed its character was its convergence with a strong premature winter storm coming down from Canada. Notwithstanding the above, hurricanes in general should not be taken lightly. Even the mildest category one's can kill if somebody is in the wrong place at the wrong time. But what makes a hurricane costly and deadly has nothing to do with global warming or climate change, but rather where it hits. If the point of contact is in a highly populated area with costly structures, then it is going to be a deadly and costly hurricane. If on the other hand, a hurricane lands in a relatively desolate area, no matter how strong, there will be minimal damage and fatalities.

A third corollary of The Theory is the extinction of species, notwithstanding Charles Darwin's admonition on the subject cited in the preface. Interestingly, we are not hearing about the extinction of such species as the mosquito, termite, rat, or the many species of poisonous snakes, etc. No, the extinctions with all the press are the warm and cuddly animals, such as the giant panda, the polar bears and the penguins, all of which hold a certain charisma. One must wonder if this is not just a coincidence…

Let's first take a look at the polar bear. Who has not seen or forgotten the iconic image of the stranded mother and polar bear cub floating on a small sliver of ice, and obviously destined for a watery grave by drowning? This image from the book, <u>An Inconvenient Truth</u>, could not have been better staged by a Hollywood producer at one of our best zoos. Perish the thought! In any event, the polar bear has seemingly become a poster child for The

Theory. According to The Theory, the polar bear is a vulnerable species headed for oblivion as a result of global warming and the resultant melting of the Arctic Ice Cap. There are predictions that the ice may be completely melted as early as 2050. Since the polar bear's diet is made up of mostly seals, which are found at the edge of sea ice, its food source would be eliminated. During summer months when no sea ice is present, they rely on stored fat. If summer passed without a recurrence of sea ice (and seals), the polar bear would obviously starve.

The theorists would have us believe that the polar bear is a fragile species, defenseless against the ravages of climate change.

But how fragile and vulnerable species is the polar bear, and if indeed it is imperiled, how responsible is global warming? First, let us deal with the "poster child" polar bear on the sliver of ice.

While it is true that there was a widely circulated report of four drowned polar bears in 2004, the actual paper states that the four bears had "presumably drowned".4 The sightings were done from the air, and it is obvious that no autopsies had been conducted to determine the cause of death. The paper mentions that there was a severe storm for days before the sightings with "…*winds <on land> peaking at 54 and 46 km/hr….and winds offshore considerably higher.*"5

I am not suggesting that polar bear drownings are implausible, but isn't there at least the possibility that the cause of death was other than drowning? How about hypothermia due to oil spills coating the fur and reducing insulation[6]? And isn't there the possibility that the bears had died on land and were blown out to sea in the fierce storm? A few possibilities come to mind: Poisoning from pollution[7], wounds from mating fights[8], and maybe even old age! The same paper cited above states that sightings in the area had been conducted for seventeen years of scores of polar bears, but never prior to this report had there been a sighting of a presumably drowned individual.

The polar bear, after all, is most probably the strongest swimming land based mammal on earth, being almost at home in the water as on land. Its thick blubber affords excellent insulation from cold water and also allows it to remain buoyant in water with no effort.[9] One female polar bear, which had been tagged was recorded as having swam 426 miles in open water before reaching sea ice.[10]

But with all that said, what if the four bears had died of drowning in that storm, isn't it even more of a stretch to conclude that the cause of drowning was receding ice as a result of global warming? This is so typical of the theorist crowd. The only cause of anything must be global warming; and then when anything happens, it confirms what they believed in the first place. We will deal with the fallacy of circular arguments in a later chapter.

Then regarding their fragility, contrary to portrayal, the species is obviously quite resilient, having survived past periods of extensive deglaciation. A polar bear jawbone fossil dating back 120,000 years has been found[11], showing that polar bears have already survived two warming periods hotter than present.[12] Moreover, a 2007 report from those who would know best state that the polar bear is doing just fine. *"At present, polar bear populations are robust and, according to native people, are considerably larger than they were in previous decades."*[13]

And finally, a compelling argument can be made derived from the following fact: Although it is true that polar bear numbers reportedly have been in a decline throughout most of the twentieth century, it appears that the trend has not only stopped, but may have even reversed. What could possibly account for this turn in events in view of the decline in north polar ice sheeting, which continues to the present? Might it just be that the world wide controls on hunting polar bears are succeeding?

Yet another of the endearing Ursus genus (yes, recent DNA analysis confirms that the panda is in the bear family rather than the raccoon) animals is reportedly at risk of extinction. Now global warming or climate change or both, threatens the continued existence in the wild of these beautiful animals. What is the evidence?

Well, it would be difficult to argue that the giant panda in the wild has not been vulnerable to the threat of extinction; since the Chinese, who should know the plight of the creature better than anyone else, have taken the issue seriously in exacting severe penalties for poaching (the death penalty reduced to 20 years imprisonment in1997)[14] and setting aside over four thousand square miles[15] of land beginning in 1963 to create fifty-six sanctuary areas[16]. The rarity of the species has been well documented, with estimates of giant pandas in the wild ranging from 1600 in a 2004 survey[17] to as many as 3000 in 2006 based on DNA analysis of droppings.[18] Further supporting the contention that the species is rare is the fact that is that its existence was considered as myth in the western world until 1869, when a French missionary received a pelt from a hunter.[19]

So, we accept that the species is vulnerable to the threat of extinction; but what is the connection to global warming or climate change?

"Recent research shows that the major source of food for pandas - bamboo - will be severely affected by an increase in global temperature. Pandas spend 14 hours a day eating, with bamboo making up 99 percent of their diet. Of the more than 100 varieties of bamboo, pandas only eat about 20 of them. And a joint study by researchers in China and the UK found that rising temperatures will cause extinction for some types of bamboo. Waning food sources will threaten extinction of China's giant panda. The best way to ensure the survival of pandas is to stop global warming."[20]

Giant pandas are known to be finicky eaters; so I suppose we should believe that they will starve rather than eat other varieties of bamboo.

The IPCC has acknowledged that the main threat to the continued existence of giant pandas is loss of habitat due to intrusion by humans. China's burgeoning population, road building, forestry, and farming have all been responsible for pushing the animals into smaller areas. Poaching has also taken its toll since the soft fur is prized and brings a high price. And, of course, according to the IPCC and fellow theorists, global warming and or climate change is a threat as well.

However, similarly to conservation efforts on behalf of the polar bear, there are reports that the wild giant panda population has stabilized and may even be increasing. *"The conservation solutions to save the species are working - and, after years of decline, panda numbers are thought to be increasing."* [21] *"In 2004, a survey counted 1,600 pandas - 40% more than were thought to exist in the 1980s."* [22]

Let's now take a look at another of the charismatic species said to be doomed to extinction by anthropogenic warming, the penguin. There are far too many references to that effect to enumerate; so I will just give you a random sampling.

"Over the past 25 years, some Antarctic penguin populations have shrunk by 33 percent due to declines in winter sea-ice habitat."[23]

"Disappearing sea ice around Antarctica may put emperor penguins at risk of extinction within the next century, warn scientists writing in this week's Proceedings of the National Academy of Sciences."[24]

If any of you have seen the brilliantly produced and poignant "March of the Penguins", you do not have to be a theorist to have been very moved by the plight of these charming animals. It would appear that the penguin may be headed for the same fate as the dinosaur. But again, what is the connection to global warming? After viewing the film twice, I came away with the feeling that the Emperor Penguin species faces some daunting challenges for survival; but it seems that the species has survived, not because it is so far from a warm climate, but in spite of the rigors of such a cold climate.

After the female lays her egg, the male in a gender role reversal, takes the egg and protects it from the cold while the female takes the long trip back to the water to obtain food for itself as well as the chick to be. When the transfer of the egg from female to male is not completed within seconds or with enough care, the egg is exposed to the frigid air and begins to crack. At that point the chick is lost. Also some of the weaker females die along the trek and some fall prey to ocean predators when they do make it to the water. Some of the males abandon their charge after they give up waiting for their returning female. Others, remaining loyal to the end, will die of starvation, awaiting their mate who will not return. In each of these cases, the chick perishes as well.[25]

A very different perspective is gained from a report that indicates that while one subspecies of penguin is declining in numbers, another is flourishing and in numbers far exceeding the decline. *"Adélie penguins have decreased by 22% whereas Chinstrap penguins have increased by more than 400% over the past 25 years (Fraser and Patterson, 1997; Smith et al., 1999). This pattern supports the hypothesis that the increasing availability of open water as a result of warmer winters is favoring the survival of Chinstraps over the ice-dependent Adélies (see Fraser et al., 1992)."*[26]

Moreover, it appears that the very method of determining penguin population trends is influencing the trends. *"...data <has been> collected from penguins marked with flipper bands...to predict impacts on the entire population from climate change scenarios modeled for the future. Flipper bands are metal markers attached to the upper part of the front flipper,* *where they are easily visible on land or in the water. Since penguins are birds that fly through the water, these "flippers" are actually wings and thus the sole source of a penguin's swimming power. Banding involves a single loop slipped over the upper, muscular section of one wing … and is meant to remain there for the life of the bird. But might these bands have an influence on the data being collected on these penguins?...*[27]

As it turns out, those flipper bands do indeed affect the data. A ten year study conducted to answer that very question, found that banding *"impairs both survival and reproduction, ultimately affecting population growth rate." The ten year study revealed that banded birds were found to produce 39% fewer chicks and had a survival rate 16% lower than non-banded birds."* [28] Talk about a self-fulfilling prophesy!

Also contrary to popular opinion, the penguin is hardly an exclusively cold climate bird or a fragile species. Fossils of early penguins have been found dating back sixty to seventy million years, and there is a species of penguin existing today that makes its home close to the equator.[29]

Regarding the other threatened species, the list is too long to go into detail on each; but according to the IPCC, whales, seals, dolphins, wolves, gorillas, rhinoceroses, elephants, tigers, etc., are all threatened by anthropogenic global warming. And the warnings go on with ever more species at risk. Scientists predict that global warming could contribute to the mass extinction of wild animals in the near future. One report states that as many as one million of the world's species may be extinct by 2050.[30] When one examines the evidence on many of the species indicated, such as the elephants and the giant pandas being pushed into smaller and smaller areas and the poaching of elephants for their ivory, the poaching of tigers and polar bears for their skins, *etc.*, it appears that the theorists may be half right on this corollary after all when they say that the danger is from anthropogenic warming. The danger may indeed be anthropogenic! The warming part is what I question.

And finally, there are predictions that the extinction of species may even extend to homo sapiens. Dr. Frank Fenner, a ninety-five year old retired microbiologist from Australia offers the following: *"Homo sapiens will become extinct, perhaps within 100 years<due to overpopulation and continued greenhouse gas emissions.>"*[31] And Prince Charles, whose scientific credentials are likely as impressive as Al Gore's, concurs in a warning reminiscent of the ancient prophets, that if we don't mend our wicked ways mankind is doomed. *"We are, of course, witnessing what some people call the sixth great extinction event – the continued erosion of much of the Earth's vital biodiversity...<and> without the biodiversity that is so threatened, we won't be able to survive ourselves."*[32]

The above is a good lead in to the last of the corollaries, i.e. that global warming will adversely affect human health. As with the other corollaries, there is no lack of studies, reports, and/or warnings. This corollary holds that global warming is responsible for increasing incidence of disease, heat stroke, deaths from hunger and thirst, drownings, *etc.* It seems that the only fatal maladies not attributed to global warming are those from hypothermia. But with more of the conversation shifting from global warming to "climate change", just wait; hypothermia may yet be included. The logic being, that global warming causes climate change and that climate change may indeed cause isolated cold spots.

Of all the fatal illnesses ascribed to global warming, heat related illnesses, including heat stroke, cardiovascular and pulmonary events triggered by high temperature would seem to have the closest correlation. It may be arguable as to whether the earth is still in a warming spell and if so to what degree; but there is no argument on the deadliness of heat waves. Severe heat waves do indeed kill, and of the fatalities attributed to severe weather (whether or not severe weather is increasing or are a result of global warming), heat waves lead the list.[33] Between 1988 and 2010 in the United States, more than 3300 deaths were attributed to heat waves.[34]

In 2003 over 37,000 lives were lost in Western Europe due to a prolonged heat wave[35], and in 2010 a heat wave in Russia claimed 56,000 lives.[36] But what about death from cold waves? We have already stated that heat waves lead the list over other severe weather occurrences in number of fatalities. Blizzards are included, but interestingly, cold waves do not seem to rate high enough to be considered severe weather. So, is there mortality data on hypothermia excluding blizzards? Indeed there is and what it shows is that cold deaths far outnumber heat deaths. [37] Is it unfair to ask if heat deaths are more regrettable than cold deaths?

Moreover, isn't it reasonable to assume that as global temperatures increase, cold deaths will decrease? At least one study indicates that this may be a fact. *"The results indicate that for most of the cities included, global climate change (global warming) is likely to lead to a reduction in mortality rates due to decreasing winter mortality."38* And lastly, isn't it likely that the mortality numbers are skewed towards the hot side by the fact that most of the world's population live in warmer latitudes? [39]

Disease is another of the scourges associated with global warming Gore states that *"some 30 so-called new diseases have emerged over the last 25 to 30 years <as of 2006>. And some old diseases that had been under control are now surging again."[40]*

Then Gore goes on to list twelve of those diseases with microphotographs of the associated pathogen.41 In his movie, he lists fifteen diseases, ostensibly included in the 30 so-called new diseases. And a report by The Wildlife Conservation Society published in National Geographic in 2008 alludes to a "deadly dozen" that may be aggravated by global warming.[41]

We generally associate diseases such as Malaria, Yellow Fever, West Nile Fever, and cholera with the tropics; so it would be natural to tie disease to hotter climates. But as we know, things are not generally as simple as they may seem at first blush. Flu is one of the exceptions that come to mind. Since disease comes in so many different forms let's first begin our examination by breaking them down by class.

Infectious disease may be classified by pathogen or causative agent, such as bacteria, virus, protozoa, parasites, *etc.* or by mode of transmission, which for our purpose is more illuminating. Pathogens may be transmitted in a number of ways, including airborne, direct contact, through blood, and ingestion.

We will start by examining the blood route since so many of the diseases are carried by mosquitoes, one of the favorite vectors of the theorists. And if mosquitoes are a favored vector, Malaria has to top the list for the most mentioned disease associated with global warming. Contrary to popular opinion, however, Malaria is not an exclusively warm climate disease. Dr. Paul Reiter, a world renowned expert in Malaria and mosquito-borne pathogens, points out that Malaria was endemic in England in the Shakespearian era, which was during the "Little Ice Age", when global temperatures were considerably cooler than present.[42] Moreover, it was endemic in most of the contiguous US, including the state of Washington and as far north as the New England states and eastern Canada until the 1940's, when the CDC was founded, and a concerted preventive campaign began.[43] Furthermore, most of Europe, including Russia has encountered malaria outbreaks in the last century, enough so that it was considered a major health issue in a number of countries.[44] And perhaps, most surprisingly, history records indigenous malaria as far north as Finland in the nineteenth century.[45]

Dr. Reiter has refuted the notion that the disease has a simple correlation to temperature or climate. To the contrary, he pointed out that climate is only one of several factors that interact to create a favorable situation for the spread of the disease.

He used the analogy of chess to characterize the complexity of malaria: *"Malaria is ... a thousand different diseases and epidemiological puzzles. Like chess, it is played with a few pieces, but is capable of an infinite variety of situations."*[46] He acknowledged that temperature, rainfall, and humidity are important elements, but looking at those alone significantly distorts the true picture.

Among the factors affecting the spread of the disease frequently ignored, he cited *"four factors that are key to the transmission and epidemiology of the disease: the ecology and behaviour of humans, and the ecology and behaviour of the vectors. .. a fifth <is> the immunity of the host and of the host population."*.[47]

While we are discussing mosquitoes, yellow fever, another "tropical" disease is often mentioned as spreading due to global warming and or climate change. Again, history records and repeated epidemics in a number of US seaports and as far north as Boston in nineteenth century.[48] Theorists may argue that these epidemics were not indigenous. That all diseases originate somewhere and then are transplanted elsewhere; and over adapt to become indigenous to the new area. More importantly, Death does not between an indigenous epidemic and a transplanted one.

Yet another of the mosquito carried diseases on the theorist's list is West Nile Virus, named for where it was first detected in Egypt in 1937 although it is thought to have first made its appearance in Greece and has been speculated to have been the disease that killed Alexander the Great.[49] Whether or not, it has emerged in the US and since its first appearance in New York in 1999 has since (as of 2010) *"...been detected almost in every province of Canada and the contiguous regions of the United States..."*[50]

While WNV differs from Malaria in the type of pathogen, i.e., Malaria is a parasite, and WNV is a virus, they are both carried by mosquitoes; so a comparison of sorts is valid. Paul Reiter had characterized Malaria as chess-like in its complexity. West Nile Virus in comparison must make Malaria look like checkers, since WNV adds another piece to the game, i.e., birds. Birds serve as an amplifying host, increasing the strength of the virus to the point where it can infect humans.[51] A mosquito bite from an infected mosquito which has not previously bitten a bird will be ineffectual in a human. Further complicating the situation is that different bird species vary in their amplification. Also the range of birds serves to spread the virus.

Another of the highly complex diseases is plague, a disease mainly transmitted to humans by fleas residing on rats. Increasing its complexity is the fact that the plague bacteria may also be transmitted directly from rat to human through bites or ingestion of food contaminated by rat excrement. Moreover, it may be transmitted from human to human when the disease infects the lung and becomes pneumonic plague.

One of the deadliest and most terrifying diseases in human history, plague, is high on the theorists' list. Never mind that the Black Death occurred mostly in the temperate areas of Europe during the 1300's and 1400's when the Little Ice Age was in progress. Also we should disregard the facts presented in a medical review, pointing out that climate change and climate factors impact host, carrier, and pathogen and each other in a complex manner that defies simplistic conclusions.

"... human plague outbreaks in several African countries were less frequent when the weather was too hot (>27°C) or cold (<15°C) . Subsequent studies showed an increased plague incidence in Vietnam during the hot, dry season, <followed by> a period of high seasonal rainfall.

"...Flea development rates increase with temperature until they reach a critical value; then the survival of immature stages decreases if high temperatures are combined with low humidity... Conversely,.. excessively wet conditions in rodent burrows ...can promote the growth of destructive fungi that diminish larval and egg survival.

"...Rodent survival <is> also affected by climate. A direct effect occurs when high intensity rainfall causes flooding of rodent burrows.

"...Human activities and behavior in plague-infected areas <including poor sanitation and hygiene as well as improper disposal of garbage> are also to be considered as important determinants of plague transmission to humans.

In the US Southwest *"... above normal precipitation in winter and spring was used to explain increases in human plague cases, and high summer temperatures, decreases of its incidence in the same area.."* [52]

One of the more curious diseases prominently mentioned in association with global warming is Lyme disease, a disease transmitted by ticks. Hardly a tropical disease, its name derives from where it was first discovered, Lyme, Connecticut. Maybe it was during one of those extremely hot summers. In any event since it first appeared, it has traveled west and south, not north.[53]

Of the diseases transmitted by ingestion, cholera is likely the most well-known. It too is high on the theorists' list of those diseases that are increasing in spread and virology due to global warming. After all, as the argument goes, the bacterium thrives in warm brackish water and as temperatures increase; the pathogen will multiply and spread. Unfortunately, for the argument, cholera is no respecter of climate, as history shows. Indeed, it is thought to have originated in the subcontinent of India in ancient times; but severe epidemics have periodically occurred in Europe as far north as Russia and in North America into Canada.[54]

Four diseases that show up on most theorists' list are smallpox, SARS, tuberculosis, and influenza. Although the tuberculosis pathogen is a bacterium and the others are viruses, they may be grouped together because of their main route of transmission. All are spread from person to person, mainly by inhaling aerosolized droplets containing the pathogen from the infected individual coughing. Any correlation to global warming is scant or nil; and in fact, a case could be made for the converse, since colder weather results in people staying indoors and coming in closer contact with one another. Moreover, smallpox flourishes in cool dry weather and is inhibited by hot humid conditions.[55]

And finally, what may be the biggest disconnect of all with regard to the deleterious effects on health due to global warming is that of malnutrition and hunger. Most of us first learned about photosynthesis in elementary school, i.e., green plants use CO_2 in conjunction with water to grow; and without either, die. Moreover, intuition, would tell us that green plants are better off with more CO_2 than less (assuming adequate water); and studies indicate that is so. Not only do plants grow faster and larger, they are healthier and more able to ward off disease and insect infestations.[56]

Before we leave this chapter, I would like to share an interesting observation. It seems that since 1998, when global temperatures leveled off, there has been a subtle shift in emphasis by the theorists from "global warming" to "climate change". Perhaps there is a corollary I have missed. Namely, global warming causes climate change; ergo, evidence of climate change proves global warming. But wait a minute! Wouldn't global cooling likewise cause climate change?

Part II
Inconvenient Statistics

"Figures don't lie, but liars figure."

Anonymous

CHAPTER 8
CORRELATIONS

Correlation: *"The simultaneous change in value of two numerically valued random variables: the positive correlation between cigarette smoking and the incidence of lung cancer; the negative correlation between age and normal vision."*[1]

In the previous chapter on corollaries we dealt with the supposed correlation between global warming and certain catastrophic events. In this chapter we will explain the difference between a true correlation and a fallacious one mathematically. The simplest correlation is a linear correlation, between two variables, which we will term x and y where a percentage change in x is accompanied by a percentage change in y; and where the ratio of the percentage change for x to the percentage change for y remains constant. This does not imply that the percentage changes must be equal, but only that the percentage changes remain fixed in their relationship.

To illustrate: $x = 2y$ This means that when x changes value, y changes by 1/2. This is a perfect positive linear correlation. Perfect because the relationship never changes. Positive because the change in x means a change in y in the same direction. If the equation is $x = -2y$, the correlation is now negative, but it is still a perfect linear correlation. Non-linear correlations occur as well, such as $x^2 = 2y^3$ but their existence normally may be determined only by complex mathematics or computer.

In the real world, there are no perfect correlations. Only in the laboratory and the heavens is there to be found a near perfect synchronicity. The reason is that there are innumerable variables, not just the x and y of the above illustration. Some are known, but many more are unknown and impact on the situation in unpredictable ways. Variations of the following graph[2] from the Vostok ice core data have been used by the theorists to show an apparent correlation between atmospheric CO_2 levels and global temperatures.

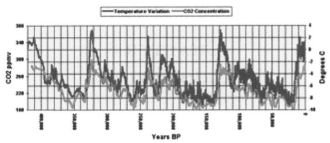

Antarctic Ice Core Data 1

Note on graph presentation: The heavier temperature lines 160,000 BP to present reflect more data points for this time period, not necessarily greater temperature variability.

When we see a graph of two lines representing two distinct phenomena moving together in such remarkable synchronicity over eons, skepticism is natural and warranted. In this situation, there is not just an x and y, but countless other variables, many of which we mentioned in previous chapters. Among the other variables are solar forcing, other greenhouse gases (mainly water, which is a far stronger greenhouse gas than CO_2), the various positive and negative feedbacks, and, the convection effect.

Moreover, as was pointed out in chapter 5, the CO_2 peaks are severely understated. But with all that said, it must be admitted that there appears to be an uncanny correlation between atmospheric CO_2 levels and surface temperatures. So are we conceding that the theorists' main argument is correct? Well, not exactly. To borrow a phrase from the late Paul Harvey, here is "the rest of the story."

There is another characteristic of correlations that should be mentioned and that is cause and effect. In the previous mathematical equations when x varies, y follows. But if we change the value of y, x then follows. Outside pure mathematics, there are correlations where changes in x result in changes in y, but not the reverse. A classic example is found in the quote from Adam Smith in Wealth of Nations: *"It is not the multitude of ale-houses . . . that occasions a general disposition to drunkenness among the common people; but that disposition, arising from other causes, necessarily gives employment to a multitude of ale-houses."*

Although it is difficult to see from the above graph because of the compressed time periods, the scientists involved in the study state that the temperature peaks precede the CO_2 peaks by about 800 years. . "*...ice core studies have shown that CO_2 starts to rise about 800 years (600-1000 years) after Antarctic temperature during glacial terminations. These terminations are pronounced warming periods that mark the ends of the ice ages....*" [3]

If increasing atmospheric CO_2 levels are the cause of higher temperatures, how can it be that the temperature rise comes earlier? Is this some kind of nature clairvoyance? Isn't it more reasonable to believe that changes in temperature are the cause of changes in CO_2 levels rather than the other way around? When Gore was questioned about that, he stated that it was a complex issue. Theorists maintain that it only goes to show that the feedback is real.

That obscures the issue, however. There is no debate on whether or not there is a feedback between atmospheric CO_2 and global temperature. The issues here are which one initiates the sequence and is there a correlation between the two. The theorists may indeed again be half right. It seems that there is a correlation, albeit imperfect; but they appear to have confused effect with cause.

Chapter 9
Anomalies

Anomaly - A deviation from the norm

The theorists must love the word, anomaly. They use it to explain anything that deviates from the Theory. In fact, one might infer that anomalies are part of the Theory as often as they are used. If there is a particularly frigid winter, that is an anomaly. If there are a number of years when global temperatures do not increase, that is an anomaly. And they even use the word to depict temperatures, both globally and for any given location that are in agreement with the Theory.

Anomaly is one of the more interesting and clever contrivances of the theorists to construct a timeline of global temperatures. It adroitly evades the issues of unavailability of temperature data from time and place; and at the same time presents a simple and convincing argument in favor of The Theory. To clarify what they are doing and see whether you agree if it makes sense, let's use an analogy:

Say, a city school system superintendent is trying to make a case that under his leadership, standardized test scores for high school seniors have increased significantly over the last ten years. There is only one problem, however. Over the last ten years, there have been a number of annexations at various times, and none of these annexed areas used the same standardized test. As it turns out, each of these annexed suburbs had higher achieving students than the city schools. If he just averages the test scores for all seniors each year, the result will be gratifying since the average will be skewed upward with each new annexed suburb. But this would be an unacceptable approach since the bias would be plain to see. So instead of using absolute test scores, he decides to use test score anomalies. The deviation from the norm will be standardized and supposedly compensate for any bias as new systems are incorporated.

Now he needs to design a system for computing the test score norm for each of the annexed areas. The simplest would be to just use the initial test score as the norm, but how do we know that the initial test score was either a positive or negative aberration.

If he instead uses the average of the test scores for each area, any effect of an aberration for a particular year would be minimized. Let's now look at the table below and view the results.

A Really Inconvenient Truth

Year	1995	1996	1997	1998	1999	2000	2001	2002	2003	2004	Ave
# HS seniors in system	5000	4900	5200	5150	5525	5917	6875	6325	6620	5884	
# HS seniors in city	5000	4900	4900	4850	4825	4800	4775	4750	2900	2800	
Ave. test score in city	60	59	59	59.9	60	62	60	61.5	59.5	59	59..99
City anomaly	0.01	-0.99	-0.99	-0.09	0.01	2.01	0.01	1.51	-0.49	-0.99	
# HS seniors in annex A			300	300	200	612	200	300	620	622	
Ave. test score in annex A			82	78	80	75	78	79	80	90	80.25
Annex A anomaly			1.75	-2.25	-0.25	-5.25	-2.25	-1.25	-0.25	9.75	
# HS seniors in annex B					500	505	700	825	1300	1300	
Ave. test score in annex B					82	70	81	84	86	87	81.7
Annex B anomaly					0.33	-11.7	-0.667	2.333	4.33	5.33	
# of HS seniors annex C							1200	450	300	412	
Ave. test score in annex C							93	84	81	89	86.75
Annex C anomaly							6.25	-2.75	-5.75	2.25	
# of HS seniors annex D									2000	750	
Ave. test score in annex D									95	88	91.5
Annex D anomaly									3.5	-3.5	
Ave. weighted anomaly	0.01	-0.99	-0.83	-0.22	0.03	0.09	0.96	1.18	1.3	1.45	

Using the weighted average test score for each area as its norm, the average weighted anomaly shows a steadily increasing trend starting at year 2; so that is the method that the superintendent elects to use. However, the astute reader may conclude that I had purposefully designed the example to yield my desired result by manipulating the inputs. Really? That is the whole point; that by using this technique one can obtain any result one wishes. All you have to do is be selective in your sampling and devise a system for determining each point's norm.

Chapter 10
Records and Probabilities

Theorists are fond of pointing out record high temperatures to support their contention that our planet is in a continuing warming trend, and Al Gore in An Inconvenient Truth cites a number of record highs in the US during that very hot summer in 2005. To prove I am a good sport, I will add a few record highs that he may have overlooked: Ozark Arkansas 120^0, Millsboro Delaware 110^0, Monticello Florida 109^0, Pahala Hawaii 100^0, Orofino Idaho 118^0, Collegeville Indiana 116^0, Keokuk Iowa 118^0, Alton (near) Kansas 121^0, Greensburg Kentucky 114^0, Plain Dealing Louisiana 114^0, Cumberland and Frederick Maryland 1090, Mio Michigan 112^0, Moorhead Minnesota 114^0, Holly Springs Mississippi 115^0, Medicine Lake Montana 117^0, Minden Nebraska 118^0, Runyon New Jersey 110^0, Steele North Dakota 121^0, Gallipolis (near) Ohio 113^0, Phoenixville Pennsylvania 111^0, Perryville Tennessee 113^0, Seymour Texas 120^0, Martinsburg West Virginia 112^0, Wisconsin Dells Wisconsin 114^0

OK, maybe I'm not such a good sport after all. The dates from the above record high temperatures were not in 2005. They happen to have been set in the 1930's decade. Moreover, not only were these record highs for these individual towns; they were and are the record highs for their respective states.[1] You might be wondering about the other twenty-six states. Surely there were some statewide records set in that sweltering summer of 2005. No, I hate to disappoint. There was one in 2006 but the most recent statewide high before that was in 1996. Below is a table of statewide highs by decade. (updated last in 2006).[2]

2000	1990	1980	1970	1960	1950	1940	1930	1920	1910	1900	1890	1880
1	6	2	2	1	4	0	23	2	5	0	3	1

Just in case you are curious, the table below shows low temperature records by decade.[3]

2010	1990	1980	1970	1960	1950	1940	1930	1920	1910	1900	1890
1	8	5	4	3	2	3	11	1	3	5	4

I anticipate objections from certain theorists like our friend, Professor Bill Chameides , whom we met in chapter 3, who may argue that the US is too small a part of the global surface to place much significance on local records. So, I checked high temperature records for each of the continents and what do they say? The most recent high temperature record was set in Antarctica in 1974, and all the rest were prior to 1950.[4]

So, what significance should we attach to record temperatures, or for that matter, record rainfalls, record droughts, and record "you name it"? It is not certain when or where the first recorded temperature was , but one thing we can be certain of is that it set a number of records, as in both, the highest and lowest temperature ever recorded, not only for that locality, but for that country, continent, and world..

The first American college football game is said to have been played in 1869; although that game resembled rugby and soccer more than our current American football. Again, the winning team set a scoring record in that game, not only for an individual team, but for all of American college football. Likewise, the losing team set a record as well, for the fewest points scored. As more teams were added and more games played, there were more records set in accordance with probabilities. Complicating the record books were the changes in rules and scoring over the years. Not until 1912 was six points allotted for touchdowns; so is it fair to compare college football scores since with those early days?

Likewise, is it fair to compare global temperature anomalies when there have been so many changes over the years? Or for that matter, individual station temperature anomalies, when the station is in an urban area that has experienced tremendous population growth, leading to the distorting "urban island effect".

Notwithstanding the above, for there to be a day, month, year or decade with multiple high record temperatures in the US or the world, should not be surprising, given the vast number of points where temperatures are recorded. Nevertheless, one can expect that any time record high temperatures occur; the theorists will trumpet such occurrences as irrefutable evidence of The Theory, and occasionally when there is an abnormal cold spell or snowfall, we will hear that they merely confirm that climate change is occurring. In the spirit of fairness, it should be noted that many dissenters will point to record cold temperatures or snowfalls as evidence to the contrary, although you won't find many dissenters pointing to record hot temperatures as evidence of global cooling.

CHAPTER 11
MISLEADING CHARTS

Charts and graphs are frequently used in statistics to display data in a simple and much more digestible format. Occasionally, however, that very simplicity facilitates manipulation of the data if that is the author's intent.

Take a look at the following graph:

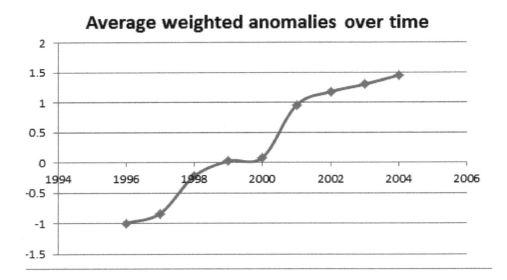

You probably will not recognize the above graph, but it represents the anomaly data from our example in chapter 9 on increasing test scores in a city school system. Pretty impressive except you may notice that the year 1995 is not shown. The superintendent obviously considered that first year as irrelevant information.

As discussed in the third chapter, the "hockey stick" curve pictured below has been used to support the contention that global temperatures are significantly higher than at any time in the last 1000 years, and of greater importance, have been rising at an accelerated rate since the mid 1800's when the industrial revolution began.

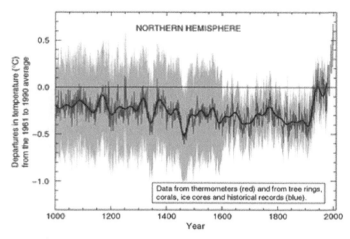

I have a question for you. Based on this graph, what is the approximate percentage increase in the northern hemisphere temperature from the low point around 1900 to the end of the graph in 1998? I understand that it is probably impossible to discern exact points in the graph; so I will make the problem easier. The low point appears to be about -0.5 and the high about +0.7. So thinking back to grade school arithmetic, we would divide the increase by the base and multiply by 100%. We obtain an answer of -240%, but that can't be right since the temperature has increased; so let's just ignore the sign and say the percentage increase is 240%. Really?

Maybe instead of using anomalies for the computation, we should use actual temperatures. Since we really do not know what the average global temperature is now, or was in 1900; nor do we know the average northern hemisphere temperature; we will use the temperature 59^0 F, which is the temperature usually cited as what our planet is at as a result of the greenhouse effect. Since Mann's graph is in degrees C, we will convert. 59^0 F equates to 15^0C (degrees C equals degrees (F-32) X5/9).

Let us then use 15^0C as the average temperature of the northern hemisphere in 1900. Based on that, the average temperature in 1998 would have been 16.20C. Doing the arithmetic, we get 8%. That appears to be reasonable, but let's check our answer using degrees Fahrenheit.

A rise of 1.2^0 C equates to a rise of 2.16^0 F. Doing the arithmetic again, we come up with 3.66%. What is going on here? Shouldn't the percentage increases be equal regardless of scale? Well, yes.

What is the percentage increase in temperature from 20' C to 22'C? 10%? Not quite; try 2/293=0.68%. Let me show you why. Take a look at going from 1' C to 3'C. That would appear to be a 200% increase. Why should a 2 degree change make such a big difference? The reason is that Centigrade (or Celsius) and Fahrenheit are arbitrary scales, based on certain physical phenomena. To calculate percentage increases we must use the absolute temperature scale called Kelvin, which is based on the lowest possible temperature. 0 K equates to -2730 C or -4590 F. If you go from the freezing point to the boiling point of water, in Centigrade it would be 0 to 100 and in Fahrenheit 32 to 212, the percentage increase is the same. 00 C equals 273 K. 100^0 C equals 373 K. Doing the arithmetic, the percentage increase is 36.6. In the case of Fahrenheit, the starting point is 491 (32 +459). If we divide the increase of 1800 by 491 and multiply by 100 we again get 36.6%.

So back to our question: What is the approximate percentage increase in the northern hemisphere temperature from the low point around 1900 to the end of the graph in 1998? With our starting point being defined as 150 C, the corresponding Kelvin temperature is 2880. The increase is 1.20, so the percentage increase for the 98 years is a whopping 0.42%! Not exactly readily apparent from the graph.

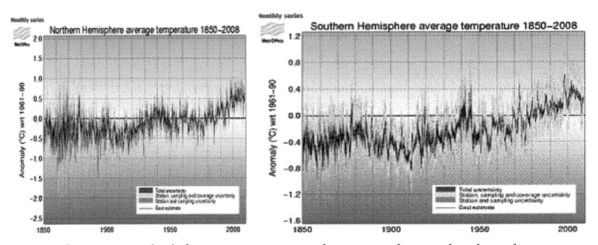

Another caveat to look for in comparing graphs is expanding or shrinking the measurements in one axis to either exaggerate or minimize a relationship. At first glance, the above graphs, originating from the Hadley Center in the UK,[1] comparing hemispheric warming look almost like copies of one another, but look closer and you will see that the temperature y- axis scales are different. Why change the scales when the data could easily have been fitted on graphs with the same scales. For that matter why not present both hemispheres on the same graph with different colors to facilitate understanding? I have my own suspicion, but I will leave it to you to arrive at your own conclusion.

Chapter 12
Statistical Validity

Before we get into the meat of the subject, let's return briefly to Dr. Mann's Hockey Stick curve. We have already seen that the curve is egregiously deceptive, but there is something else about it that should be mentioned. Simply stated, it has been determined to be statistically invalid!

When Steve McIntire, a Canadian geologist and expert statistician, first saw the curve, he found it peculiarly similar to many he had seen in his own field, i.e., gold mining, where unscrupulous mining promoters used deceptive graphics to tout highly speculative ventures based on non-random sampling techniques. McIntyre had neither a background in climate nor any axe to grind in the global warming debate, but his interest was stirred.

He began e-mailing Mann, requesting the underlying data, and after hounding Mann tirelessly, McIntyre was finally able to obtain the data he needed to check Mann's results. What he found confirmed his earlier suspicions of fallacious statistics. According to McIntyre, s analysis, Mann had used a mathematical program that would elicit a "hockey stick" curve 99% of the time using random input. When the same data was processed according to acceptable statistical practice, however, the "hockey stick" disappeared.[1]

After McIntyre posted his results, a firestorm of controversy erupted. After all, the Hockey Stick curve had purportedly shown that northern hemispheric temperatures were at millennial highs for the decade of the ninety's and 1998 the hottest year. The curve had been featured prominently in IPCC's climate report of 2001 as was the statement on millennial high temperatures, which buttressed Theorists claims that global temperatures had been rising since the industrial revolution, tied to CO_2 emissions

Considering the serious economic ramifications for the US as well as the world, Congress entered the picture and requested Dr. Edward Wegman, a renowned statistician with impeccable credentials to assemble a team to perform an inquiry. He agreed to take on the project pro bono and asked several other top notch statisticians to join him. When they finished they concurred with McIntyre's assessment. This repudiation of the curve led to the IPCC removing any reference to it and its conclusion in subsequent climate reports.

Back in chapter 3 when we were discussing the measurement of the earth's temperature, considering that there were far more empty cells than cells with temperature measuring stations, I made the semi-facetious remark that it might be preferable to "just use four recording stations: one at each pole and two at the equator spaced at 180^0 longitude apart."

Surely I jest when I say semi-facetious? Not really. Which is more absurd, having thousands of temperature measuring stations concentrated in certain areas, but leaving vast areas of our planet totally unmonitored, and if that is not enough, arbitrarily adding and subtracting test stations as well as altering historic data over the time period in question; or my semi-facetious suggestion? Which method has a higher degree of statistical validity? It is intuitive that numbers of samples alone don't assure validity. The selection method is obviously as important.

Let's assume that a particular city has a serious domestic abuse problem, (which for the most part goes unreported) in an area of that city, with a population 25% of that in the entire city; and for purposes of this illustration that percentage has remained constant for ten years. If we were wanting to ascertain the rate of domestic abuse increase or decrease in the city over that ten years, rather than interview each household (particularly in a large city) we would probably do a sampling.

The correct method would be to set up a grid and make sure that 25% of the households polled would be from the high abuse area, which was done ten years earlier. But let's say we now run into a problem of finding enough volunteers to go in to this area since it is also a high crime area; so the result is that the poll for that area only covers 10% of those polled. Now we analyze the results and are pleased to find that domestic abuse in the city has gone down significantly!

I will concede that using four temperature samples (even if the locations have not changed) hardly qualifies as a statistically valid sampling. So, how many samples are needed to qualify? Let's first define what it is that we want to measure. What is meant by the average temperature of the earth? Well, at any point in time it would be the average of the infinite number of points on the earth's surface. Obviously, it is impossible to measure an infinite number of anything; so we have to strike a compromise between a practical number of measurements and small enough areas that will hopefully have little average variance in temperature. So again, how many samples are necessary? It all depends on what margin of error and confidence level one is willing to accept.

In any statistical study there is a defined margin of error in consideration of the fact that the sample is only a statistical estimate of the entire population, which in this case is the average temperature of each of the infinite number of points on the surface of the earth.

As stated in chapter 3, NOAA and HadCru have divided the earth's surface into 5184 cells, 5^0 longitude by 5^0 latitude, which vary in area as a result of converging longitudinal lines approaching the poles; whereas GISS has chosen to use 8000 cells of equal area. In 1975, the year with the greatest number of test stations[2], there were approximately 6200 stations spread unevenly among those cells. (27% were in the U.S.) There is no question that a larger number of test stations within a cell would strengthen statistical validity for that cell, assuming there are no extraneous distorting factors, such as a nearby rubbish burner or airport runway; but with respect to the whole earth, each cell may be considered a sample point and additional test stations superfluous.

In order to achieve a degree of statistical validity for a date in the past without increasing or relocating a number of stations, (which is impossible for a past date) the cell size would have to be enlarged to the point that it encompassed a minimum of one test station. But how would we determine the required size? We will have to make some assumptions; but all assumptions will be best case scenarios favoring The Theory.

The surface area of the earth is approximately 197 million square miles, and Antarctica has a surface area of about 5.1 million square miles. I chose to look at Antarctica because it is a large defined area with relatively few test stations. In 1975 there were 29 test stations (but only data for 27) located on the continent.[3] If we assume each of the 27 were spread evenly over the 5.1 million square miles, cell size would be approximately 189,000 square miles with one test station in the middle of each cell. If we then use 189,000 square miles as the cell size over the surface of the earth (in order to have an equivalent area in each cell), there would be 1043 cells, each containing a minimum of one test station for the year 1975. We will assume that all cells have their test station locations averaged in the center for best possible validity. We will further assume that the recorded temperature for a cell with one or a small number of test stations is equivalent to the average temperature over the surface of the cell, and we will assume that the average temperature of all of the cells outside Antarctica is equal. Lastly, as discussed in chapter three, we will use the generally acknowledged (and estimated) average temperature of the earth to be 15°C (59°F).

What then is the margin of error with the above assumptions? The formula for margin of error (MOE) with a 95% confidence level is as follows:

MOE= 1.96σ Sigma (σ) is the symbol for standard deviation of the population, but since the population standard deviation is unknown we will approximate it by using the sample standard deviation (s) given by the following formula:

$$s = \sqrt{\frac{\sum (x - \bar{x})^2}{n - 1}}$$

lowing table will show the temperatures and calculations for each of the Antarctic well as for all cells outside Antarctica.

		A	B	C	D
		1975 Antarctic Stations4	Ave.Temp °C	x-15	(x-15)²
1		Faraday	-3.2	-18.2	331.24
2		Base Orcadas	-3.9	-18.9	357.21

3	Signy Island	-4.4	-19.4	376.36
4	Almirante Brown	-2.3	-17.3	299.29
5	Bellingshause	-3.25	-18.25	333.0625
6	Centro Met An Marsh	-1.2	-16.2	262.44
7	Base Arturo P	-3	-18	324
8	Matienzo	-13.7	-28.7	823.69
9	Cms_vice. Do.Marambio	-8.7	-23.7	561.69
10	Base Esperanz	-6.8	-21.8	475.24
11	Bernado O'Higgins	-3.7	-18.7	349.69
12	Mimyi	-10.9	-25.9	670.81
13	Dumont D'Urvi	-10.8	-25.8	665.64
14	Mawson	-11.1	-26.1	681.21
15	Molodeznaja	-11.1	-26.1	681.21
16	Davis	-10.5	-25.5	650.25
17	Syowa	-11.25	-26.25	689.0625
18	Leningradskaya	-15.1	-30.1	906.01
19	S.A.N.A.E. St	-14	-29	841
20	Novolazarevsk	-10.1	-25.1	630.01
21	Halley	-17.8	-32.8	1075.84
22	Scott Base	-19.8	-34.8	1211.04
23	McMurdo	-16.3	-31.3	979.69
24	Belgrano	-22	-37	1369
25	Belgrano_i	-21.9	-36.9	1361.61
26	Vostok	-55.6	-70.6	4984.36
27	Amundsen-Scot	-49	-64	4096
28	Totals for Antarctica cells	-361.4	-766.4	25986.66
29	All cells outside Antarctica	1016X	1016(X-15)	1016(X-15)2

Explanation of calculations in table and following are in notes.[5]

$(1016X-361.4)/1043= 15$

1016X-361.4=15*1043= 15645
1016X= 15645+361.4= 16006.4
X= 15.75°C
X-15= 0.75
(X-15)2= 0.56
1016*(X-15)2 = 569
25986.66+ 569= 26555.66
n-1 = 1042
26555.66/1042=25.49
Sq. root of 25.49= 5.05
MOE (95% confidence) = 5.05* 1.96=9.90

Recalling that the earth is said to have warmed 0.8° C since 1880[6], it is interesting to say the least that the above statistical analysis indicates a margin of error of ± 9.90°C. How can the Theorists be so certain, or in the words of Mark Twain, "...*know for sure*..." that the global surface temperature is accurate?

Notwithstanding the above, let's take a look at one more year to gain some perspective. There were only seven stations in Antarctica (six with 1950 data) up until 1950.[7] Using the same math as above, each of the six cells would cover 850,000 square miles and over the surface of the earth, there would be 232 cells.

	A	B	C	D
	1950 Antarctic Stations	Ave.Temp0C	x-15	(x-15)2
1	Faraday	-3.1	-18.1	327.61
2	Base Orcadas	-6.1	-21.1	445.21
3	Signy Island	-3.9	-18.9	357.21
4	Base Esperanz	ND		
5	Dumont D'Urvi	-12.1	-27.1	734.41
6	Deception Island	-3.8	-18.8	353.44
7	Rothero Point	-4.8	-19.8	392.04
8	Totals for Antarctica cells	-33.8	-123.8	2609.92
9	All cells outside Antarctica	226X		

(226X-33.8)/232=15
226X-33.8= 15*232= 3480
226X= 3480+33.8 =3513.8
X= 3513.8/226=15.55
X-15= 0.55
$(X-15)2= 0.30$
226* 0.30= 67.8
2609.92 + 67.8=2677.72
N-1= 231
2677.72/231=11.59
Sq. root 11.59= 3.40
MOE (95% confidence) =1.96*3.40 = 6.67

There may be some surprise that the margin of error is smaller with a lot fewer samples. The explanation is simple. In 1950 and years earlier there were no measurements in the coldest areas of Antarctica, resulting in a lower variance. If we wanted to reduce the margin of error even further, we could use Faraday as the single station and cell for the entire 5.1 square miles of Antarctica. Using the same assumptions and math, I calculate a margin of error of 4.8^0.

In fact, I could make the margin of error all but disappear, or like Merlin make it disappear completely with the right selection of test stations and extrapolation of the data over the surface of the earth. Does that sound familiar? Of course, the Theorists will condemn my statistical analysis as totally invalid at best, and perhaps an excursion into fantasy land at worst, given all of the assumptions. I would reply that most of the invalidity stemmed from my leaning over backwards to accommodate the Theory, but how much further into fantasy land would we have traveled if not for all the assumptions that minimized the margin of error?

In view of the above statistical analyses, the use of anomalies to manufacture historical data, and changing numbers and locations of test stations over the last century, would it be unreasonable to conclude that the entire surface temperature record is statistically invalid? Considering the great stock that the theorists place in consensus, I will leave it to the Court of Public Opinion (which I might point out includes a far greater number of learned scientists who are dissenters than theorists [8]) to decide.

Part III
Inconvenient Economics

"You can't have your cake and eat it too."
Anonymous

CHAPTER 13
EXPLAINING ECONOMICS

I think it may be safe to state that most physical scientists have never had a course in economics; and while the same may not be said of the members of Congress, which is made up of mostly lawyers, who generally have undergraduate degrees in the social sciences, at times it seems that the only economics they understand or are interested in has to do with their continued congressional paychecks and pensions.

And whether or not economics courses show up in their transcripts, it is probably a better bet that most scientists, physical and social, have never run a business larger than a lemonade stand. For the rest of us, whether we had an economics course or not, it is probably time for a refresher.

Economics is sometimes referred to as the science of choices or priorities. All resources are limited (although many politicians appear to disbelieve that fact), forcing individuals, families, organizations, and governments to choose where to spend those resources.

In many cases it appears we have but two choices: Do we really want that new car or do we want to put that money away for our child's education? These days a medium priced car is one year's tuition and board at one of the better universities. But economics is not as simple as that. Our choices have consequences, many of which are hidden or not apparent.

Children are not good at establishing priorities; a child may well decide to spend his allowance on candy bars instead of apples or other more nutritious foods, even if he or she has been told that candy results in unpleasant visits to the dentist's office. But adults are often no more proficient in making choices. Let's go back to that car or college decision. If our present car has seen its better days and is a gas and oil guzzler, and we have a long daily commute; it might be best to consider other options, e.g., a low priced new car, a used car, or maybe even a motorcycle. Perhaps we might even encourage our youngster to start earning some money by mowing lawns or babysitting. Things are just not that simple.

When it comes to energy, whether or not we subscribe to The Theory; we may still feel it is prudent to reduce our energy consumption for conservation purposes or just to save money. We are not lacking for choices: We can purchase one of the fully electric cars, a hybrid, or downsize to a motorcycle. We can car pool, forego a car trip to any destination less than a few blocks, and instead walk or ride a bicycle. We can have our car retrofitted to burn biodiesel.

We can get rid of all our incandescent bulbs and convert to the low wattage mercury bulbs. We can raise our thermostats in the summer to 80^0 F and lower them in the winter to 60^0F. And on and on.

Some of those choices may be delusional and others may result in consequences that actually increase costs or use of energy. The fully electric car will cost more and may only be worth it from a financial standpoint if we drive enough miles and keep the car long enough. Trade-in value at this early stage is a big question mark. And what if we lose our job and the only available job in our field is further away than the range of our vehicle? Time to either move or trade in for a gas powered car.

It is obvious that decision making is complicated by unforeseen developments and our inability to see into the future. Of the many alleged threats to humanity and our planet's ecosystems from global warming, the one threat with an undisputed correlation to global warming is a rising sea level. As the planet warms, there will be significant melting and a significant sea level rise. But how much and in what time frame? In Mr. Gore's movie, "An Inconvenient Truth", he states that the sea level could rise twenty feet in the foreseeable future. As he did not define "foreseeable future", it is difficult to argue the point, since he may feel that he can foresee into the future further than most of us. Nevertheless, the IPCC in their 1990 Impacts Assessment predicted a sea level rise of "up to one meter (a little more than three feet) by 2100." This was revised in the 2007 Impacts Assessment to 0.28 to 0.43 meter by 2100.[1]

Even this amount would be catastrophic to large numbers of third world coastal residents, forcing relocations as well as eliminating land currently used for growing crops. Even in developed countries such a sea level rise would adversely affect the poor by infiltrating low lying river deltas with saline water. Wealthy coastal areas would not be exempt from impact, but classic economics would come into play to resolve the problem. If the asset value exceeded the cost of protection (levees), protection would be the choice. For everyone else, assuming that there is no reversal in global warming (and that is a big assumption, as we shall see in a later chapter), what should be done about this threat? What are our options?

Three basic options present themselves:

1. We could work to drastically curb fossil fuel use by enacting laws to penalize carbon emissions and encourage alternative energy solutions (more about that in the next chapter).

2. Considering that there has already been about a one foot sea level rise since 1860 without major disruption[2], and that the projected rise amounts to little more than a foot; we could utilize our resources to add protection as needed to the vulnerable areas.

3. Or we could take a middle road and hedge our bet by slightly reducing carbon emissions, while also adding protection to affected areas.

An intriguing study based on IPCC data and assumptions concludes that the best approach by far is option #2; and counter-intuitively, taking the middle road actually results in a poorer outcome with more flooding at higher cost because the cost of implementing even modest reductions reduces economic growth and the available resources to provide protection.[3]

"…*More than a foot of sea-level rise will cause about one-hundred million people to get flooded each year <if we do nothing>… <however>, with smart protection at fairly low cost it is likely that we will see not one hundred million people flooded each year but fewer than one million…Today we have ten million people getting flooded- eighty years from now, with higher sea levels and more global warming and many more people, we will not see this number increased tenfold, but actually decreased more than tenfold, because of rich societies being able to deal with flooding much more effectively.*"[4]

CHAPTER 14
ALTERNATE ENERGY SOURCES

If fossil fuels are the bane of our world, how are we supposed to replace them for our energy requirements? Alternate (renewable) energy is the answer, say the Theorists. But, is it? There is an apparent common misconception that anyone who takes issue with The Theory is adamantly opposed to all forms of non-fossil fuel energy, with the exception of nuclear energy. What we dissenters strongly object to, however, are the onerous taxes and penalties on CO_2 emissions, coupled with government incentives (taxpayer money) that are being proposed to encourage "green energy". If there exist now or in the future any alternate technologies that can stand on their own competitively, we say bring them on. The fact is, however, that the renewable energy sources being promoted are either non-competitive or have other limitations that restrict their use currently to relatively minor amounts of energy.

Solar power is the darling of the Theorists, and admittedly has intriguing possibilities in the future; but on the other hand, presents an exquisite irony in that Theorists appear to discount its influence on whatever warming may be occurring. Although dramatic advances in technology have lowered the cost per watt of solar cells to about parity with coal in the last two or three years, there are several problems. The first is expanding from the single cell to a solar panel increases cost while reducing efficiency. Then putting the panels together in an array adds additional cost and further reduces efficiency. What may be the biggest drawback is the amount of land required for a solar energy plant. If one looks only at the amount of solar radiation hitting the earth, the number is intriguing: 1,368 watts per square meter. This would translate to 1.37 megawatts per 1000 square meters of solar panels. But this does not take into account days when the sun is partially or totally blocked by clouds, and it also assumes that the entire spectrum of radiation is converted to electrical power. I did a search on the internet and the best rating on a commercial solar panel that I was able to find was 16.2 watts per square foot[1], which translates to about 174 watts per square meter. Keep in mind that this number is based on continuous sunshine unimpeded by clouds.

If solar is the darling, wind power may rate a close second, although not with nearly as unanimous favor for reasons we shall see. However, similar to solar, wind power is free for the taking, can be used by individual homeowners (assuming no problems with association covenants or government ordinances), and best of all emits none of that detestable CO_2. Moreover, wind turbines can be placed on farm lands without detracting from agricultural production. Power density of course varies based on wind speed, making it impossible to assign a single value for a given turbine; but since available energy is proportional to wind speed cubed, wind energy can reach very high levels in severe storms. At a height of 50 meters, a wind speed of 26.6 miles per hour equates to 2000 watts per square meter.[2] Since energy is proportional to wind speed cubed, doubling the wind speed would multiply the wattage by eight. And tripling the wind speed would multiply the wattage by twenty-seven. Wow!

Unfortunately, there are other factors that adversely affect usable energy from a wind turbine. First, because wind cannot leave a turbine at zero velocity, the maximum theoretical energy from a wind turbine is 59% of the total energy. Further losses result from rotor blade and gear box friction as well as generator and converter inefficiencies.[3]

Second, there are two wind conditions that result in zero energy output, very low wind speed and very high wind speed. Wind turbines need a minimum speed to "kick in", normally about five miles per hour and they are set to cut off at very high wind speeds to prevent burning up the motor. And that formula for converting wind speed to energy works in reverse as well, so when the wind speed drops from maximum output by 1/2, wattage is reduced by 7/8.

Those preceding factors result in an actual realized energy output which is termed load factor. Again, because wind conditions make up a significant part of load factor for a wind turbine, load factors continuously change and also vary by location. A five year government study in the UK found that average load factors were 26.2% for on shore wind turbines and only slightly higher for off shore wind turrbines.[4] Average load factors in Germany, Netherlands, and Denmark are said to be even lower at between 19 and 22%.[5] Moreover, a recent study found that wind turbines were wearing out much faster than anticipated, losing as much as one third of their load factors within ten years.[6] The load factor when multiplied by a wind turbine's energy rating gives the true yield.

Then there is the issue of area requirements. Since the wind coming out of a turbine has been slowed, it would be senseless to place another one too close. So how close is close? Spacing requirements differ based primarily on the configuration, which in turn depends on prevailing wind conditions and terrain. According to the US National Renewable Energy Laboratory, if prevailing winds are multidirectional, turbines should be spaced five to seven rotor diameters apart. Closer spacing of three to four rotor diameters perpendicular to prevailing winds is warranted when winds are generally along the same line.[7]

As of this writing (2012) the largest wind turbine in the world is the Enercon E-126 with a rated capacity of 7.58 MW.[8] Based on a rotor diameter of 126 meters, and a power factor of 25% it has an actual operating capacity of 1.895 MW, which computes to 152 watts per square meter.[9] But power density in a wind farm containing a large number of turbines decreases that significantly as we shall see. For some perspective let's consider a local coal burning steam plant with a rated capacity of 2840 MW[10] and an estimated load factor of .42, based on 2010 data for power capacity and consumption for all US electrical generating plants.[11] This computes to an actual operating capacity of 1192.8 MW.

To match the capacity of that plant, it would take approximately 629 Enercon E-126's. Now let's examine land requirements. The above steam plant occupies less than 300 acres[12] (1,214,057 sq. meters). Using the closest spacing of five diameters from the above NREL guideline for a multidirectional wind and a square configuration each E-126 in a wind farm containing 625 (for simplification) E-126's would require an average of 335,937 square meters[13] or less than 6 watts per square meter. The wind farm would occupy 209,960,604 square meters or over 50,000 acres.

So what's the problem? There is plenty of unused land and coastal waters around the world. Some of us may recall idyllic scenes in the past of a farm with a single windmill; but a wind farm with scores of giant wind turbines does not often evoke sentimental memories. To many they are an eyesore and a blight on the environment.[14]

In addition to the land requirement, there are reports that the low frequency sound emitted by wind turbines may be harmful to health.[15] Moreover, conservationists are pointing to massive bird kills.[16] There have been protests in Canada, Australia, Wales, and the US just to cite a few. And these protesters are by no means all Theory dissenters.

Another of the alternative and renewable energies, like wind power that has been around for a long time and is "free for the taking" is hydropower. Hydropower taps the kinetic energy of falling water by the use of dams, which let the water through in a controlled manner to turn turbines. This gives hydropower the flexibility to ramp up and down quickly in response to changing energy demand, which is an advantage that wind power lacks. In addition to flexibility, a major advantage is low energy cost, despite high capital costs, which can be amortized in a relatively short time due to low operating and maintenance costs. The dams may provide flood control as well as aiding agriculture through irrigation. They also may offer an attractive side benefit with lakes for recreation and tourism.

But again, notwithstanding all the benefits, there are some substantial detriments; and similar to wind power, there are detractors in the ranks of the Theorists. Dams may result in injury to aquatic ecosystems; and the reservoirs created by dams submerges vast areas of land, which may cause relocations of large numbers of people. Also the submerged areas often include grasslands and forests as well as farms, which then offsets any irrigation advantages downstream Submerged grasslands and forests create another problem by destroying plants that remove CO_2 through photosynthesis, and even far more important, subjects the submerged plants to anaerobic decay, which produces methane, one of the strongest greenhouse gasses in contrast to CO_2, which is the weakest.

And then there is an alternative energy source which most of us in the US are using regardless of whether we want to or not. I am referring, of course, to ethanol for helping to power our automobiles. Never mind that ethanol when combusted emits that same nasty CO_2 as fossil fuels. And non-scientists may be forgiven for their ignorance of the fact that the manufacture of ethanol from corn, sugar or other bio-feedstocks involves a fermentation process that also emits CO_2.[17] They may also be ignorant of the fact that the product of fermentation is mostly water, which must be removed from the ethanol before it can be used as a fuel. This purification process is generally distillation, which requires energy input of some kind, whether from fossil fuels or alternative energy sources. Let me think... How about ethanol? Sorry, I just couldn't resist..

But, not to worry, the Theorists are always ready for a rejoinder: *"Although the burning of biomass also produces carbon dioxide, the primary greenhouse gas, it is considered to be part of the natural carbon cycle of the earth. The plants take up carbon dioxide from the air while they are growing and then return it to the air when they are burned, thereby causing no net increase."*[18] That in my estimation is one of the most asinine, if not the most asinine, arguments of the entire Theory, and ironically plays right into an argument refuting the Theory. Where do they think the fossil fuels got their carbon in the first place? Furthermore, I was always under the impression that green plants love and consume CO_2, regardless of the source.

One university study makes a very valid point in that the best energy use for biomass is burning for fuel[19], rather than converting to another energy source, which requires energy.

Notwithstanding all of the above, I say again if there is an alternate energy source that can stand on its own without taxpayer assistance, bring it on. Some Theorists may protest, accusing me of hypocrisy considering that petroleum, natural gas, coal, and uranium companies all receive government subsidies in the form of depletion allowances. Those Theorists would only be displaying their ignorance of economics. The depletion allowances are tax credits, which reduce the tax bill and can occur only if the company has earnings. If the company has no earnings it would owe no taxes and therefore receive no tax credit. I think a better approach than tax credits for any company is to totally eliminate corporate tax, since they amount to a double tax on shareholders, who are taxed on dividends, but that will have to wait for another book. Until corporate taxes are eliminated, I for one see no problem in offering similar tax credits to alternate energy companies; but this will not save companies like Solyndra that despite taxpayer money could not break even, let alone make a profit on which they would owe taxes.

CHAPTER 15
HOT VS. COLD

According to the IPCC, based on our present trajectory, atmospheric CO_2 levels are forecast to be between 550 and 950 ppm by 2100.[1] Also, IPCC models show that global temperatures in 2100 will rise accordingly 1.1°C to 6.4°C (2⁰F to 11.5 ⁰F) .[2] So in a worst case scenario, three or four generations from now, it would be a much hotter and more uncomfortable world. Before anyone panics, however, he or she might want to re-read the chapters on measuring earth's temperature and statistic validity, and then decide if maybe there aren't more important things to be concerned about.

Which would you rather be, hot or cold? My guess is that if you posed that question to a large group of people, you would get a variety of responses, including variations of *"Neither"* or *"May I get another choice?"* or *"How hot or cold?"* I suspect that most of us are like Goldilocks, *"Not too hot or cold- just right."* But relative humidity needs to be factored in as well. On a very humid day, a temperature that at normal humidity would be delightful, would feel intolerable. Moreover, *"just right"* varies between individuals in an area; and there is even a wider variance when the individuals are from widely different climates. Typically, preferred temperatures are somewhere in the mid-range of high and low temperatures for that particular climate.

A different ambient temperature parameter far more important than the above preferred temperature and with intriguing ramifications has been referred to as the optimal temperature for humans. Optimal temperature is that temperature at which human mortality is lowest.[3] If temperatures are hotter or colder than that temperature, deaths increase; but interestingly, the optimal temperature, like the preferred temperature above varies by location, and has been found to be approximately the average summer time temperature for that location. As one would expect then, warmer climates do have higher optimal temperatures and cooler climates lower. But another interesting fact is that the optimal temperature changes over time as the climate changes. As the climate warms, the optimal temperature goes up and vice versa.

Let's play devil's advocate and assume that IPCC worst case scenario holds true. A 6.4⁰C increase sounds like a disaster and sure enough the average increase would not be the same for the entire planet. The bad news is that land would be warmer and oceans cooler relatively because water takes more energy to change temperature. The good news is that colder latitudes would warm more and hotter latitudes less. Moreover, winter temperatures would increase more and summer less. Lastly, night time temperatures would warm more and day time less.[4] All of these effects would lead to a moderation of temperatures over the earth.

We prefer moderation of temperature enough that we are willing to pay for it in the form of air conditioning in our homes and cars, central heat; and for those who can afford it a winter or summer house. Some others of us may be able actually to afford that second home; but forego it, because we have other priorities. So temperature is another of those economic choices.

Indeed, global warming has a price. The IPCC has estimated that in a "business as usual" scenario with no effort to control global warming through 2100, the cost to the global economy would amount to $14.5 trillion.[5] So, with that kind of cost it would only seem logical to spend some money now to curtail the warming. Logical as it would appear, it turns out that the cost to make a significant impact would be several times that of not doing anything about the threat. A proposal by the EU to hold the global temperature increase to less than 2^0C has been estimated to cost the world economy 84 trillion or almost six times the cost of doing nothing.[6] Hardly a good deal.

Some Theorists might acknowledge those economic forecasts but question my ethical priorities. Am I not placing greater importance on financial considerations than the welfare of the world and its creatures? I don't think so. A result of unwise financial decisions will be a poorer world less able to deal with the needs that allegedly will come with the remainder of global warming that we have not curtailed. Less money for hospitals and medical research; less money for flood and drought control with dams, irrigation and water desalinization, less money for animal sanctuaries, less money for air conditioning, etc.

For some perspective on that 6.40 C increase, let's look at average 2011 temperatures (the last temperature data available as of this writing) for two US test station locations:

Bismarck ND 4.8^0C

Phoenix AZ 24.3^0C

A difference of 19.5^0 C! Obviously, I could have shown an even wider difference by comparing locations in northern Canada to those in Mexico; but the point is not that an average increase of 6.4^0C is insignificant. Rather that humans are adaptable to widely varying temperatures. If by some sort of magic in a perverse grand experiment by climatologists to prove the devastation of climate change, the populations of Bismarck and Phoenix were instantly transported to each others' cities, there would indeed be a lot of misery, ill health, and probably fatalities in both locations.

But even the most ardent theorists are not predicting such a change in the near term. Rather the forecast is to occur over this century, spanning three or four generations with plenty of time for adaptation. Remember too this was the high end estimate. So the question is how much is it worth to you to mitigate a 4.5^0 temperature increase for your grandchildren or great grandchildren? Remember too that the cost will not only come out of your pocket, but the same people you are trying to protect.

In 2004 a panel of economists including four Nobel laureates was asked to rate comparative strategies and chances of achieving success for dealing with some of the world's most pressing problems and opportunities.[7] There were seventeen choices ranging from disease, malnutrition, and sanitation to climate. The choices with the poorest chance of success: Mitigating global warming! A follow up panel with five Nobel laureates was convened in 2008, and this time there were thirty choices. The two choices that came in last and second to last? Reducing CO_2 emissions and R&D into reducing CO_2 emissions.[8]

Part IV
Inconvenient Facts

"Facts are stubborn things; and whatever may be our wishes, our inclinations, or the dictates of our passion, they cannot alter the state of facts and evidence."

John Adams, 'Argument in Defense of the Soldiers in the Boston Massacre Trials,' December 1770

CHAPTER 16
MIS-STATING FACTS

Of all the misrepresentations of fact, including those already discussed in chapter five, and easily refuted, the statement that the issue is settled is probably the most pervasive. Close behind is the notion that very few scientists who have expertise in climatology do not accept The Theory.

According to the theorists, 2500 scientists from around the world, who are the expert contributing authors and reviewers for the IPCC concur in the notion that the evidence for human caused global warming is undeniable and that unless drastic and immediate action is taken to reduce CO_2 emissions, global catastrophe lies ahead in the form of droughts, flooding, disease, species extinction, etc. Also are the numerous scientific organizations such as the Union of Concerned Scientists and the American Meteorological Society, who echo their unanimous support. According to the theorists, the few scientists that deny any part of the above are either being paid off by fossil fuel interests or they are incompetent crackpots.

Politicians of most developed countries have bought in to this dogma; and the media has jumped in to spread the word. Al Gore has enlisted an army of volunteers that travel to Nashville at their own expense to attend a training program, which schools them in Theory dogma and supplies them with an elaborate slide show presentation that they are happy to present to any church, organization or school interested in being "educated" on the impending disaster of global warming.

This onslaught has been very effective in convincing large numbers of otherwise intelligent people with no scientific background that The Theory is gospel. With that deluge of information, how otherwise is an intelligent lay person expected to separate fact from supposition or even fiction? Shouldn't you listen to and heed the advice of the experts, particularly when an overwhelming consensus says it is so?

Well to begin with, as Dr. Fred Singer reminds us in his essay, "Global Warming: Man-Made or Natural?"[1], science is determined by the scientific method, not by a show of hands. Einstein had said despite all the scientists who had accepted his theory of relativity, it would take only one experiment by one scientist to disprove it. And ironically, Dr. James Hansen, Director of NASA's Goddard Institute for Space Studies and one of the leading theorists, states in his book, <u>Storms of My Grandchildren</u>, that his "… *favorite description of the scientific method..*"[2] is the following quote by Richard Feynman: *"The only way to have real success* *in science...is to describe the evidence<for the theory> very carefully without regard to the way you feel it should be. ...you must try to explain what's good about it and what's bad about it equally..."* I would like to ask Dr. Hansen if he had actually believed in the truth of that statement at some earlier time, or if he has been a hypocrite all along.

As for the consensus noted above, the debate is far from over, and the supposed consensus is a myth. First, for the latest ICCP report (2007) only 600 of the 2500 scientists participated in the preparation and review of the 1000 page first section, termed WG-1, which concerns the **possibility** (my bold) of humans having an influence on global warming.[3] The most crucial chapter in WG-1 and underpinning for The Theory is chapter 9, which removes the conjecture and states that humans are the primary cause of climate change, was authored and reviewed by only less than one-half percent of the 2500.[4]

Despite the IPCC's statement that there would be a wide diversity of viewpoints among the contributors in order to insure objectivity, the opposite was the case. In John McLean's incisive and meticulously researched analysis he exposed the creation of an illusion of a consensus through the use of a "stacked deck" of participants that included co-authorships as well as colleagues and subordinates of the same institution.[5] When two of seven unbiased expert reviewers disputed the conclusion that humans had primary responsibility for any perceived effect on global warming or climate change, their comments were ignored and not included in the final report.[6] Any bets on whether these individuals will be invited back for the next assessment?

Likewise, only a small number of the governing board of the American Meteorological Society voted on the "consensus statement" regarding climate change while the rank and file members were never consulted, leading to open rebellion by many. *"Estimates of skepticism within the AMS regarding man-made global warming are well over 50 percent."*[7]

Also, as of this date (October 12, 2012) 31,487 American scientists with educations in fields of specialization that suitably qualify them to evaluate the research data, including 9029 with PhD's have signed a petition stating that there is no credible scientific basis for the claim that anthropogenic release of CO_2 or other greenhouse gases has caused or will cause catastrophic warming or climate disruption in the foreseeable future. It further states that *"... substantial scientific evidence <indicates> that increases in atmospheric carbon dioxide produce many beneficial effects..."*[8] One scientist, whose name may surprise you expressed his skepticism regarding The Theory in an interview in 1998 shortly before his death. The dissenting scientist? None other than Al Gore's mentor, Professor Roger Revelle.[9]

And as one might imagine, dissenting scientists are not limited to the US. Hundreds more worldwide have joined in their dissent.[10]

In addition, there are countless scientists that although they have not publicly announced their opposition to The Theory are "closet" dissenters due to intimidation through threats of loss of funds, firing, and character assassination. How do I know this? Just look at a few reports of the intimidation that is going on to quiet any scientist that would dare to question the consensus.

At a climate change conference held in NYC in 2008 featuring skeptical international scientists, a number of US meteorologists declined to attend, fearing that if they did, their employment could be jeopardized. [11] Dr. Nathan Paldor, an Israeli professor of Dynamical Meteorology and Physical Oceanography, and climate skeptic, stated that many of his colleagues share his views, but have been intimidated by an inability to have their skeptical papers published.[12] Dr. Reginald Newell, an MIT professor, had fears confirmed that his funding would be cut after co-authoring a paper that concluded that global marine air temperature from 1888 to 1988 increased a miniscule 0.24⁰C and that apparent 20th century temperature increases represent a natural recovery from cooling in the early 1900's probably due to volcanic activity.[13] Dr. Joanne Simpson, an eminent recently retired atmospheric scientist (2007), formerly of the NASA Goddard Space Flight Center and presently the Chief Scientist for Meteorology, Earth Sciences Directorate, stated that now that she had left NASA and receives no funding, she is free to express her skeptical views.[14] She did not come right out and say she had been intimidated, but the implication is clear. George Taylor, retired Oregon State Climatologist, was forced from his position for his contrarian views on climate change, which conflicted with his state's declared position of human caused global warming.[15] Dr. William J.R. Alexander, a South African professor of civil and bio-systems engineering and former member of IPCC's scientific and technical committee on natural disasters stated that he was *"...subjected to vilification tactics at the time*<of declaring his skeptical stance>. *I persisted. Now, at long last, my persistence has been rewarded...There is no believable evidence to support [the IPCC]."*[16]

As regards the payoff allegation, ironically, the real money is on the side of the Theorists with governments, foundations, and corporations providing the funding. The US government through our taxes is most likely the largest single financial source for The Theory, spending almost $6 billion per year on climate research, which is more than is spent for cancer or AIDS..[17]

Christopher Horner in Red Hot Lies rhetorically asks his readers to consider how long they think the money would continue to flow if tomorrow scientists agreed that there was conclusive evidence that any anthropogenic warming was within natural variance and effects on climate change non-catastrophic. And author Michael Crichton in his State of Fear, summarized the situation very well, *"...Scientists are only too aware of whom they are working for. Those who fund research--whether a drug company, a government agency, or an environmental organization--always have a particular outcome in mind... Scientists know that continued funding depends on delivering the results the funders desire..."*[18]

Al Gore, although not a scientist, is more than likely the single largest beneficiary of The Theory. In early 2001, Gore's finances were reported as $7 million, just seven years later in 2008, he was worth at least $35 million,[19] a 400% increase. Dr. Hansen, his scientific advisor, has been doing very well himself. He received the Heinz Foundation Award in 2001 for $250,000 and shared the Dan David $1 million prize with two other individuals in 2007. How much Hansen has received in his role as scientific advisor to Gore is unknown.

Two other falsehoods spread by the theorists are closely related in that they allow their outrage to be focused towards a single entity. The first is that the fossil fuel industry is a monolith with regard to the belief that CO_2 emissions are not a problem. Several facts give lie to this premise, considering that a number of petroleum companies are actively researching alternative energy sources, and spending money to inform the public that they are doing so. *"Five of the world's largest oil companies -- BP PLC, Chevron Corp., ConocoPhillips Co., Royal Dutch Shell PLC and Total SA -- have given more than $550 million in the past two years* <as of 2009> *to university research programs"* for research into alternative energy; and over the last ten years (since 2002), the largest of all, Exxon-Mobil, has contributed $100 million dollars to Stanford University's Global Climate Research Project for research into alternative energy sources *"... focused on developing technologies to lower greenhouse gas emissions."*[20] Moreover, the coal industry is working feverishly to sequester CO_2; and the natural gas industry is promoting their product as the "clean" fossil fuel alternative, since its CO_2 byproduct is far lower than any other fossil fuel on an energy basis.

The second falsehood, and discussed in the introduction, is that "deniers" are a homogeneous lot, denying every aspect of The Theory, including the fact that CO_2 a greenhouse gas. Not only does such a statement improve the theorist's focus, it disparages dissenters and makes them appear anti-science. As pointed out in the introduction that statement is bogus also.

Finally, if there were any doubt that many of the misstatements of fact were inadvertent, we have the following candid admissions somewhat akin to jailhouse confessions, where inmates freely discuss their crimes with other inmates when they think the authorities are not listening, to put those doubts to rest.

There is an old adage, *"The squeaky wheel gets the grease."* In terms of the global warming debate, it goes something like this, *"He who screams the loudest and creates the most horrifying scenario, gets the most money for climate research."* Apparently, Dr. Stephen Schneider, an outspoken theorist and prophet of man-made climate warming, (years after being a prophet of man-made climate cooling) subscribed to this technique as he stated bluntly: *"On the one hand, as scientists we are ethically bound... to tell the truth, the whole truth, and nothing but... on the other hand,...we need to capture the public's imagination...So we have to offer up scary scenarios, make simplified, dramatic statements,...* <with> *little mention of any doubts we might have... Each of us has to decide the right balance between being effective and being honest".*[21]

In fairness, the quote ends with Schneider hoping that he *"can be both"*<effective and honest>. "But Sir, what happens if there is a conflict between effectiveness and honesty? Which do you choose?" The desire to be effective, despite doubts implies an agenda, which is contrary to the quest for truth, the very essence of science.

In 1998, Canada's Minister of the Environment, Christine Stewart, pretty much summed up IPCC's stance on anthropogenic global warming when she stated *"No matter if the science is all phony, there are collateral environmental benefits....Climate change [provides] the greatest chance to bring about justice and equality in the world."*[22]

Although you may not have heard of it before now, there is a corollary to the squeaky wheel adage. It goes something like this: *"If there is a second squeaky wheel, it too will require grease. Unless that one is eliminated, it will result in less grease for the first."* In testimony before a congressional committee in 2006, Dr. David Deming, a professor of geology at the University of Arizona, " *...received an astonishing email* <eleven or so years ago> *from a major researcher in the area of climate change"* stating, *"We have got to get rid of the 'Medieval Warm Period'"*[23]. Whether or not, Dr. Mann received a similar message, he apparently shared the same sentiments in the construction of the Hockey Stick graph. Apparently, the Medieval Warm Period is another annoying squeaky wheel. The sender of the email must have mistakenly thought Deming was a fellow theorist, based on a paper he had written in 1995, reviewing bore hole data that indicated a temperature increase of one degree Celsius in North America over the previous one hundred to one hundred fifty years.

The email was later attributed to Dr. Jonathan Overpeck, also a University of Arizona professor and an IPCC lead author, who couldn't recall ever sending such an email, but if he did, it was certainly "... taking the quote out of context..." [24] Years later when asked, Deming said that he had deleted the email at the time and could not positively identify the sender as Jonathan Overpeck, but Deming did say it was from *"an Overpeck"*. [25] So, in fairness, I am asking the other Overpeck, who is or was a *"major researcher in the area of climate change"* to do the honorable thing and step forward to reclaim his email and thereby exonerate Jonathan Overpeck.

Chapter 17
Ignoring Facts

When a courtroom witness is sworn in, the witness is asked not only to swear to tell "nothing but the truth", but also to tell the "whole truth". There is good reason for this, as the omission of certain pertinent facts may be as misleading as an outright lie. "Cherry picking" or "data mining" may be used to support a position when the ignored data actually contradicts a conclusion.

We have seen how dependent The Theory is on proxies for filling in the climate record prior to the time of instrumented measurements. Fair enough, but is it really fair to exclude those proxies that are in direct contradiction, such as plant stomata, which was discussed in the chapter on proxies? Referring to them as outliers or aberrations, and then ignoring them, is at best naive, and at worst, dishonest.

Moving to the present day, we find that NASA satellite data shows little or no global warming as contrasted with the surface record, which we have already shown, is flawed. In an interesting irony, our friend, Dr. James Hansen, who I may remind you heads the Goddard Space Center of NASA, ignores the satellite data in saying that any suspension in the warming trend is only temporary and that the long term trend remains intact.[1] But, sir, what about the contradiction of the surface record for the same years since 1979 when satellite data first became available? Doesn't that fact call into question the validity of the entire surface record and the supposed long term trend? Isn't there reason to believe the satellite data is more accurate than surface measurement for the following reasons?

1. Elimination of bias with regard to positioning of surface temperature stations and temperature rise in growing urban areas.
2. Elimination of bias from moved surface stations.
3. Much better geographical coverage.

Along the same line, warming and cooling cycles in recent times appear to correlate well with solar activity. [2] Further, there is mounting scientific evidence that the same correlation exists over the ages! [3] But yet, the latest IPC assessment report (2007) continues to trumpet the supposed correlation between our burning of fossil fuels and downplays any role played by the sun.[4]

Another fact with regard to temperature measurement conveniently ignored is the urban heat effect previously discussed in chapter 3. The Theorists state that the urban heat effect has only minimal effect and has been accounted for in temperature adjustments. Really? How about the study by Jim Goodridge, California State Climatologist? Goodridge analyzed the California county temperature data from 1910 to 1987 and separated his readings into three classes by population. What he found was an excellent correlation between population (which corresponded to population growth) and temperature changes. For those counties having populations of more than one million there was an average temperature increase of 4^0F. Counties with populations between 100,000 and one million showed an average increase of 1^0F, and counties with less than 100,000 residents averaged 0 change.[5]

A more recent study by scientists from California State University, NASA's Jet Propulsion Laboratory and Utah State University looking at data between 1950 and 2000 from hundreds of temperature measuring stations in California confirmed Goodridge's conclusion, i.e., high population centers showed significant temperature increases; whereas rural areas showed little or no increase. In fact, *"... 41% of the stations had no significant warming and 6% actually had cooling."*[6]

Before we leave the subject, we should mention an e-mail from Dr. Madhav Khandekar, a retired environmental consultant and IPCC expert reviewer, in which he refers to a paper by Dr. Phil Jones, a prominent theorist, that touts increasing global temperatures. Khandekar states: *"...significant warming trends are present in only 10-20% of the available grid boxes... and a close look at these boxes reveal that for land- areas, most of these boxes are in the vicinity of large cities and urban centres of the world..."*[7] It would appear that if greenhouse gasses are in fact responsible for global warming, they sure are awfully picky about where they concentrate.

We continually hear the predictions of the Arctic and Greenland becoming ice free in the foreseeable future and the many glaciers around the world that are melting, but we don't hear too much about Antarctica, other than an occasional reference to the western peninsula where melting has been observed. But about the rest of Antarctica? Antarctica contains 90% of the world's ice[8] and its sea ice has actually been gaining mass at the rate of 1% per decade at least since the start of satellite observations in 1979. [9]

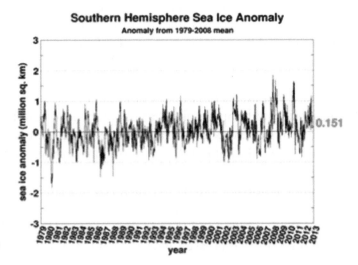

Theorists may argue that sea ice changes are not that important since they have no effect on sea level. That is true, but what about the effect on albedo? And now there are reports that show Antarctica continental ice growing as well.[10] Even more astounding is an ice core study showing that continental ice has been growing since the 1850's! [11]

Furthermore, Antarctica is not the only place where ice is increasing. We have heard the numerous reports of glaciers melting, including Arctic ice, but for some reason we have heard very little of glaciers around the world gaining mass. When one is pointed out, it is dismissed as a short term anomaly due to unusually heavy precipitation in the area, but, interestingly, a lack of precipitation is never presented as a cause for a receding glacier, as is the case with Kilimanjaro and other east African glaciers.[12] Also, it would be difficult to argue that advancing glaciers are just a regional occurrence inasmuch as they are found on five different continents.[13]

Another interesting and ironic fact ignored by the theorists has to do with aerosols, which were discussed in chapter seven. Theorists certainly acknowledge that aerosols provide a negative forcing, i.e., reducing temperature; but there is not much discussion on how aerosols are created other than from volcanoes,, forest fires, and industry. The point that is conveniently avoided is that industrial emission of aerosols comes primarily from … Are you ready for this? … the burning of fossil fuels! Any ideas on what we should call this process? Maybe something like anthropogenic global cooling?

Of all the scary aspects of The Theory, including drought, floods, famine, disease, and extinction of species, none is more terrifying than the tipping point of runaway warming, which supposedly will occur if we continue to burn fossil fuels and load the atmosphere with CO_2, which is 395 ppm as of this writing.. It's not clear what that future CO2 level is, but the Theorists assure us that we are almost there. Others say that no matter what we do, we are already on the road to oblivion.

Very interesting, indeed. There is just a slight problem. The above scenario would seem to ignore the fact that many times over the eons atmospheric CO_2 levels have been considerably above the current level.[14] Either that, or it means that we have hit the tipping point several times in the distant past. So, what happened after the tipping point was reached? Well, according to a study by Dr. Matthew Huber and associates of Purdue University, an ice age was triggered by a reduction in CO_2 levels.15 So that explains it: When the CO_2 level goes high enough to trigger a tipping point, the CO_2 level drops and we descend into an Ice Age. OK, so we don't burn up, we freeze instead. That's pretty scary, but I have a question. Where did the CO_2 go? Did the CO_2 drift off into space? Not likely, since CO_2 is denser than nitrogen and oxygen, which would go first if something caused gravity to lose its grip. Isn't it far more likely that the oceans absorbed all that CO_2? Just one more slight problem. The solubility of CO_2 goes down in hot water and goes up in cold water. So there you have it. The oceans got colder and absorbed additional CO_2 causing the planet to descend into an Ice Age through a negative feedback mechanism. So I guess global warming ultimately results in global cooling, which then brings on global warming, etc.

Chapter 18
Misleading Facts

A headline reads, "97% of climatologists say global warming is occurring and caused by humans."[1] Really? Is that a fact? Well, not exactly. The above is an example of a misleading headline. The article states that the above reference is from a survey of "*... researchers listed in the American Geological Institute's Directory of Geoscience Departments.*" The survey was sent to the 10,200 research scientists listed in the directory; but less than one-third actually responded, and of those it was 82% who felt that global warming was significantly affected by humans. 90% felt that current global temperatures are higher than temperatures in the pre-1800's. The 97% refers to active researchers among the respondents. Not too surprising considering that those doing research into global warming might not want to jeopardize their research funding by taking a negative position.

In any survey it is only prudent to check the objectivity of the source. In this case it didn't take too much digging. The authors write that the *"wide support among climatologists does not come as a surprise. They're the ones who study and publish on climate science. So I (sic) guess the take-home message is, the more you know about the field of climate science, the more you're likely to believe in global warming and humankind's contribution to it. The remaining challenge ...is how to effectively communicate this to policy makers and to a public that continues to mistakenly perceive debate among scientists."* [2]

I guess the take home message is that the more you know about the magnitude and credibility of dissent, the more you might want to mislead people into thinking that the matter is settled.

There are other ways to obscure the truth besides lying, withholding pertinent information, and covering up. By presenting information in various devious ways, telling what may be defended as the truth may turn out to be very misleading. We have seen in chapter eleven how charts and graphs may mislead.

Words and context may convey entirely opposite meanings. Sometimes just as ignoring facts may be defended as an inadvertent omission, choice of words and phrases may be defended as unintentional. Examples are Dr. Hansen testifying before a Senate committee in 1988 that he had *"99% confidence <certainty> that...Earth was being affected by human-made greenhouse gasses, and the planet had entered a period of long-term warming.."* [3] With what metric did he determine his certainty? Might he have been only 98% certain, or maybe he was 99.99% certain. He would surely defend his terminology as something everybody uses when they are absolutely certain of something…almost.

If he is absolutely certain maybe he should have said 100% certain, but he leaves some room for doubt. It reminds me of a fellow I saw the other day who was wearing a T-shirt with the following phrase: *"I may be wrong, but I highly doubt it."*

Also referring to CO_2 as the *"most important greenhouse gas"* [4][5] is almost certain to be misconstrued. We have already pointed out in chapter four, that CO_2 is far from the strongest greenhouse gas and likewise far from having the strongest greenhouse effect in our atmosphere. So what is meant by most important? Could it be that it is the most important greenhouse gas in the debate on anthropogenic warming? OK, on that one we will concede, but I think you get my gist. The choice of words, intentional or not, may be confusing to the layman.

Another example of word parsing by theorists is the following statement. *"The* [Antarctica] *peninsula is the fastest warming place in Antarctica."* That is a fact, but as Christopher Horner points out, the peninsula occupies only 2% of the continent. The other 98% is actually cooling.[6]

Dr. Hansen refers to Dr. Henrik Svensmark's theory (without mentioning him by name) on the effect of solar magnetic variations on cosmic rays and the formation of clouds as "Rube Goldberg" , because it involves complex interactions.[7] Surely, Dr. Hansen doesn't believe that climate isn't complex. But, if he thinks that the theory is not credible because of its complexity, why doesn't he have the intellectual honesty to come out and say it rather than resort to an attack that denigrates someone's viewpoint and makes it not even worthy of debate? May it be because he is afraid to have a debate on its merits?

The "denier" label attached to any person who disputes or even questions the main premises of The Theory is intended to place that individual in the same category as the "Holocaust Deniers" who are not only repugnant but crackpots as well. One environmental author even suggests future Nuremburg type trials for those who would dare to take issue with The Theory.[8]

CHAPTER 19
COVERING UP FACTS

We have already seen in a previous chapter how satellite temperature data, showing little or no change from 1979 was ignored since it conflicted with surface station data. Apparently, it was not enough for some to just ignore data that showed no global temperature increase from 1998 to 2008 based on surface station readings as well.

Roger Harrabin, an environmental reporter for the BBC News, found himself in hot water with a theorist organization after writing an article in 2008 with the headline *"Global temperatures ' to decrease'"*. Harrabin is no global warming skeptic, and the headline with the quotes around *"to decrease"* obviously intended to convey a sarcastic note. Nevertheless, his article so irritated, Jo Abess, a leader of an activist group that it prompted an exchange of several emails, which pointed out his "error" and forced him to recant and amend his article to more suit theorist doctrine. It should be pointed out that there were no factual errors in the original piece, which stated that global temperatures had not risen between 1998 and 2008 and would actually be a slight bit lower in 2008 than in 2007. The statement that this was all attributed to El Nino effects and no way indicated that there would be any deviation from the long term warming trend was only conjecture and certainly was no problem for Abess. When the article was redone it emphasized that the cooling was a temporary aberration from the long term warming trend, and although global warming skeptics would be sure to jump on this as evidence that global warming had ceased, this was not the case.[1]

Another example of temperature data cover-up and more clever was the manipulation of temperature records over the past century by NOAA. If the manipulation had been only a fudging of recent readings, it would have been too obvious. Instead under the guise of "quality control" readings from as long ago as 100 years were adjusted, not once but many times, and unsurprisingly, those distant in the past adjustments were generally downward to help with the current warming argument.[2]

And then there is Steven Schneider of "balancing honesty with effectiveness" fame, who had given a filmed interview for a documentary on global warming and candidly expressed his decades earlier belief in global cooling. He apparently had second thoughts as did his employer, Stanford University. When the university threatened to sue, *"... the documentary filmmakers were forced to use a blank screen and an actor had to read the transcript of Schneider's already taped but legally banned climate interview".*[3]

Science is based on the free flow of information not only to check facts but to help yield new insights. UK's Hadley Center apparently disagrees with this cornerstone of science. The center repeatedly denied access to a list of temperature stations around the world used in compiling their global data sets claiming the data showed no adjustment was needed.[4]

It appears that the NCDC is on the same page as the Hadley Center. After a flurry of bad press articles, accompanied often with embarrassing photographs of US test stations stemming from Anthony Watts' expose' on his web site "Watts Up With That" showing flagrant violations of siting protocol, the NCDC rushed to cover its "inadvertent" error by removing location addresses of the offending monitoring stations from the its website.[5] After much pressure from Watts and others, the NCDC relented and made the locations available. They, however, continued to resist posting station photographs that they were known to possess despite pressure.[6] They finally reconsidered and you can now view photographs of *selected* monitoring stations that surprise, surprise, fit the siting protocol exactly. One can only trust that the offending stations have not only been removed from the website, but also from processing data used in temperature calculations. Surely too, temperature records have been corrected eliminating those erroneous readings.

Finally, in what appears to be the most egregious example of fact cover-up discovered to date, a cache of more than 1000 confidential e-mails from the Hadley Climate Research Unit (CRU) of East Anglia State University in England somehow made its way into the public realm in mid-November 2009. Whether a hacker or an inside whistleblower that should be thanked, is irrelevant.

What is to the point was the content in many of the e-mails that showed data manipulation to support the hypothesis of The Theory and blatant attempts to cover up the manipulation such as the elimination of the Medieval Warm Period in the construction of the Hockey Stick curve.[7]

Part V
Inconvenient
Uncertainties

"Nothing is certain but death and taxes."

anonymous

CHAPTER 20
FUTURE UNCERTAIN

"Prediction is very difficult, especially about the future."
Niels Bohr (1885 - 1962)

Most people are aware that predicting the future is far from an exact endeavor. Some gamblers may be an exception; but most will try to hedge their bet or at least not wager all of their assets on a single outcome, which says that they know their desired outcome may not happen.

The insurance industry is predicated on the unpredictability of the future and meteorologists certainly are aware of this as shown by their forecasts. Weather forecasts are always given in probabilities. An hour or two into the future is much easier to predict than a day ahead; and forecasts normally are not given for longer than a few days ahead. Even "long range" forecasts are rarely given for more than a couple of weeks ahead.

Then, one must inquire how climatologists can be so sure of their predictions years or decades into the future. Are they that much smarter than meteorologists? I doubt that, and I think most climatologists would agree it is not a matter of intelligence. Climatology predictions are based on trends and correlations. Moreover, those predictions do not state a specific time that a given climate is to occur; only that it will or likely will occur, given a continuance of trends and certain other assumptions. The climatology Theorists tend to downplay the assumptions; but the dissenters or skeptics among the climatologists (and they far outnumber the theorists as we have seen) point out that the assumptions create so much uncertainty that it is impossible to accurately predict the future climate other than to say that at various times in the future the earth will likely be warmer and cooler than it is now and likewise dryer and wetter.

Michael Crichton, a brilliant MIT graduate and famed author of many sci-fi thrillers, including <u>Jurassic Park</u>, in a lecture on the lunacy of relying on computer models to predict far into the future, when the input involves numerous estimates and uncertainties, presented an often quoted analogy.

"Let's think back to people in 1900 in, say, New York. If they worried about people in 2000, what would they worry about? Probably 'Where would people get enough horses? And what would they do about all the horse shit?' Horse pollution was bad in 1900, think how much worse

CHAPTER 21
UNCERTAINTY IN MEASUREMENT

Every scientist knows that error is inherent in measurement. Whether one is measuring mass, time, dimension, or temperature, no matter how precise the measuring instrument, the reading is inevitably an approximation. Error may be introduced by several factors, including, the thing being measured, the measurement process, instrument problems, environment, and sampling.

With regard to the thing being measured, it may surprise most non-scientists, but there is a serious question as to whether there is even such a thing as an average global temperature. However, there are so many problems associated with global temperature measurement as to make the question moot. As was pointed out in chapter 3, the process has been flawed by a blatant disregard of NCDC protocol resulting in 69% of the monitoring stations in the US with daily margins of error of between 2^0C and 5^0C. 89% had margins of error of 1^0C or more. Given that global temperatures have purportedly gone up just 0.7^0C in the last century, it does not take a mathematical genius to deduce that something may be wrong with that estimate. Theorists may argue that the above problems only involved US test stations; so I guess we are to assume that everywhere else, including third world countries, everything was in perfect order. Interestingly, as pointed out in chapter 19, the locations are kept secret for some reason.

Environment is another contributing source of uncertainty. The theorists claim that the urban heat effect is minimal, and that effect has been compensated for. The evidence as discussed in chapter 17 just does not seem to indicate that the urban heat effect is minimal. As for the adjustment, might that refer to the creative work by Professor Wei-Chyung Wang? The good professor allegedly fabricated data for locations that have no recorded temperature monitoring stations. Moreover, of the monitoring stations that did exist many had moved multiple times.[2][3]

Sampling is yet another big source of uncertainty, given the large areas of the globe uncovered by test stations. Then compound that with the method of determining temperature anomalies for locations where there was no historic data, which was discussed in chapter 7. And then we have this comforting statement from the IPCC on converting those anomalies back to real absolute temperatures. *"The latter shows that global temperature anomalies can be converted into absolute temperature values with only a small extra uncertainty."*[4]

CHAPTER 22
UNKNOWN UNCERTAINTIES

"The Truth About The Coming Climate Catastrophe And Our Last Chance To Save Humanity" So reads the subtitle in Dr. Hansen's book. No weasel wording here. He could have said, "…the possible coming climate catastrophe…" or even *"… the probable coming climate catastrophe… "or " … and what may be our last chance…."* So, I take it he is pretty darn certain. I had earlier questioned Dr. Hansen about his 99% certainty in that Senate committee meeting, so will not beat that dead horse. However, one must assume that sometime between 1988 and 2009 (the copyright date on his book) that last lingering doubt of 1% had been resolved.

The IPCC appears to be pretty certain as well. *"Evidence is now 'unequivocal' that humans are causing global warming"* In the Synthesis Report of the IPCC's fourth and latest assessment report (2007), it states,…" Warming of the climate system is unequivocal…" *"The 100-year linear trend (1906-2005) <was> 0.74 [0.56 to 0.92]°C…"*

Theorists appear to have no doubts about their theory, but upon what do they base that certainty? Well, their many models all concur that global warming has occurred since the onset of the industrial revolution and that greenhouse gases, mainly CO_2, produced by the burning of fossil fuels, are responsible for most of that warming. Further, the models predict that if fossil fuel usage is not drastically curtailed, we are headed to a much warmer world with catastrophic consequences when we hit the tipping point, which is very close.

Over the course of this book I have enumerated countless estimates and uncertainties, including temperature estimates for missing cells in the global distribution of temperature measuring stations, utilizing proxies to estimate what temperatures and CO_2 levels were before anyone was measuring them, errors in the measurement itself as shown by the numerous adjustments made after the fact as well as siting problems with measuring stations, and then some apparently deliberate efforts to "massage" data. After all that one would think that the theorists would have more than a "warming" of one degree Celsius over 150 years to hang their hat on.

Perhaps, if they spent more of their "quality control" time actually cleaning up the corrupting influences in their temperature measurement rather than adjusting old temperature data, some might come to realize that the CO_2 warming is an illusion.

But the above are not the only uncertainties. On NASA's web site there is the candid admission that scientists know very little about some potentially strong climate affecting agents, including long term solar cycles, soot and dust particles, clouds, and others, some of which are unknown. In addition, it states that little is known about some of the effects of climate change, including precipitation and sea level rise.[1]

And yet, climate models are being used to predict all sorts of calamitous outcomes if we continue to burn fossil fuels. Can those models be accepted as credible? I leave it to you, the jury, to decide.

CONCLUSION

As I write this, we are suffering a record cold winter. Three severe snow storms have blanketed the Northeast, and the South has had unusually cold weather. Does this mean we are in for a much cooler summer or years ahead of colder and colder temperatures? I would answer that the same way I did in my introduction, I have no way of knowing. Of course, the theorists have a ready answer. It is either one of those short term aberrations from the norm, or it is just another confirmation of global warming. You see, when temperatures rise there is more evaporation, and it is well known that evaporation produces a cooling effect, so this is how warming can cause cooling.

I sure hope my readers realize that the above is "tongue in cheek", but this is the sort of convoluted logic that theorists have used to spread their "Gospel". Whatever the event, be it temperature records (high or low), severe weather, droughts, floods, famine, locusts, plague, disease, it must be a confirmation of global warming or a "short term" aberration.

When one asks what constitutes a "short term" aberration, there is a good chance that on this question there is not a "settled" answer. It probably depends on the duration of the aberration. I expect that 30 years might be an answer since this was the length of the cooling spell from the 1940's to the 1970's. But then I would ask why couldn't 90 or 150 years constitute a short term?

Certain readers may ascribe an ulterior motive to this work and others may just write it off as the irrelevant musings of another "right wing" kook. I can assure all that I am not in the employ of any company or group that would benefit if the whole "global warming" debate would finally end with the proponents acknowledging defeat. You may be surprised to learn that I happen to be for reducing our reliance on petroleum as a fuel. However, my support for petroleum fuel reduction stems not from support of The Theory, but for our country to become energy independent. As for my being a "right wing" kook, I will acknowledge that my politics tend to be moderately conservative, but more accurately moderately libertarian. I am not anti- environment, in fact far from it. I am against waste and pollution, but I am also against politicians and demagogues stirring up problems and using environmental causes for their own personal agendas and gain.

The argument is not whether there is any argument. I believe anyone who has read this book knows there is an argument. However, whether or not some warming may have occurred in the last century; and whether or not humans may have been responsible for what slight warming has occurred are not really that important. The argument is whether our world is headed to a calamity unless we take immediate and drastic action to halt CO_2 emissions.

We dissenters do not dispute that CO_2 is a greenhouse gas that contributes to warming of the planet. Most of us also do not dispute that humans have been responsible for most of the CO_2 increase since the beginning of the industrial revolution. Most of us also do not deny that humans contribute to a warming effect.

We do not agree, however, that whatever warming may have occurred, that CO_2 is a significant factor; or that further increases in atmospheric CO_2 will lead to an irrevocable increase in global temperatures, given that CO_2 is a minor greenhouse gas, and that the convection effect tempers any greenhouse effect. Moreover, predictions of gloom ignore solar effects and countless other variables, many of which scientists have not discovered, that influence our climate and have led to ice ages and warming periods over the ages.

The lack of falsifiability as annunciated by Popper is proof to any honest scientist that The Theory is pseudoscience. and the irony is that those who are labeling the true scientists deniers are themselves the deniers for denying the essence of science. The word "unequivocal" is a favorite word among theorists. It is probably an apt word, since it seems to me and hopefully, you the jury, that what is unequivocal is that the evidence supporting The Theory is far from unequivocal.

Disclaimer

Readers may come away from having read this book with the mistaken notion that I believe all scientists who are theorists are dishonest. While I do believe that there has been a lot of unscrupulous behavior, I feel that a number of scientists are sincere and may have become so infatuated with their work and egged on by all the praise that they have fallen into a type of gambler's syndrome, where a system has supposedly worked in the past and they cannot believe that it will not work in the future; so they "double down". I am not a psychologist nor do I presume to be able to read another person's mind or motive. Every scientist who falls into the category of theorist will have to examine his or her own conscience.

NOTES

Chapter 1
1. http://www.carbonify.com/articles/global-warming-hoax.htm

Chapter 3
1. GISS website
2. http://www.nicholas.duke.edu/thegreengrok/2008temps
3. http://climate.geog.udel.edu/~climate/html_pages/Global2011/GlobalTsT2011Loc.html
4. Ibid
5. http://www.warwickhughes.com/climate/az.htm
6. http://www.warwickhughes.com/climate/florida.htm
7. www.surfacestation.org
8. Ibid
9. US CRN Site Handbook paragraphs. 4.21 and 4.22
10. www.surfacestations.org
11. What Do Observational Datasets Say about Modeled. Tropospheric Temperature Trends since 1979? By Dr. John R. Christy et al www.mdpi.com/2072-4292/2/9/2148/pdf
12. http://www.appinsys.com/GlobalWarming/GW_NotGlobal.htm
13. http://www.citypopulation.de/world/Agglomerations.html

Chapter 5
1. One hundred years of Atmospheric CO2 …. By Ernst-Georg Beck
2. http://geology.gsapubs.org/content/28/4/351.abstract
3. http://www.warwickhughes.com/icecore/
4. Ibid
5. http://en.wikipedia.org/wiki/Proxy_(climate)
6. Ibid
7. http://en.wikipedia.org/wiki/Proxy_(climate)#cite_note-Boreholes_in_Glacial_Ice_p80-9

8. http://www.nap.edu/openbook.php?record_id=11676&page=80

9. http://www.seafriends.org.nz/oceano/seawater.htm

10. http://carbon-budget.geologist-1011.net/

11. http://en.wikipedia.org/wiki/Hydrothermal_vent

12. http://wattsupwiththat.com/2010/12/26/co2-ice-cores-vs-plant-stomata/

13. http://en.wikipedia.org/wiki/Murder_of_JonBen%C3%A9t_Ramsey

14. http://en.wikipedia.org/wiki/Natalee_Holloway

Chapter 6

1. http://www.atmosphere.mpg.de/enid/3rv.html

2. http://en.wikipedia.org/wiki/Albedo

3. http://www.ucsusa.org/global_warming/science_and_impacts/impacts/arctic-climate-impact.html

4. Science Daily (Nov. 10, 2009) "Antarctica Glacier Retreat Creates New Carbon Dioxide Store; Has Beneficial Impact On Climate Change"

5. Bounoua, L. et al., 2010. Quantifying the negative feedback of vegetation to greenhouse warming: A modeling approach. Geophysical Research Letters, 37, L23701, doi+10.1029/2010GL045338.

6. Ainsworth EA, et al What have we learned from 15 years of free-air CO2 enrichment (FACE)? http://www.ncbi.nlm.nih.gov/pubmed/15720649

7. http://telstar.ote.cmu.edu/environ/m3/s2/subsect/predict.htm

Chapter 7

1. http://www.nasa.gov/topics/solarsystem/features/venus-temp20110926.html

2. Balling,R.C. et al. "Analysis of tropical cyclone intensification…" http://adsabs.harvard.edu/abs/2006MAP....93...45B

3. http://www.epicdisasters.com/index.php/site/comments/the_ten_strongest_hurricanes/

4. Polar Biology Volume 29, Number 8, 681-687, DOI: 10.1007/s00300-005-0105-2

5. Ibid.

6. Stirling, Ian (1988). "What Makes a Polar Bear Tick?". Polar Bears. Ann Arbor: University of Michigan Press. ISBN 0-472-10100-5.

7. Skaare, Janneche Utne et al. (2002). "Ecological risk assessment of persistent organic pollutants in the arctic" (PDF). Toxicology 181–182: 193–197.

8. Stirling,. "Reproduction" op. cit.

9. http://www.marinebio.net/marinescience/04benthon/arcplrbr.htm

10. http://www.adn.com/2011/02/06/v-printer/1687857/polar-bears-epic-swim-seen-as.html

11. . Jonathan Amos, "Ancient Polar Bear Jawbone Found," BBC News, December 10, 2007.

12. http://muller.lbl.gov/pages/IceAgeBook/history_of_climate.html (Chapter 1) (Figure 1-2 Climate of the last 2400 years

13. Don Martin, "Polar Bear Numbers up but Rescue Continues," National Post (Canada), March 6, 2007.

14. http://www.wwfchina.org/english/pandacentral/htm/wwf_at_work/panda_survey/q&a.htm

15. http://www.chinahighlights.com/giant-panda/habitat.htm

16. Corwyn, Jeff "Giant Pandas" 100 Heartbeats, Rodale, Inc. ISBN-13 978-1-60529-847-4, ISBN-10 1-60529-847-4

17. http://wwf.panda.org/what_we_do/endangered_species/giant_panda/panda/how_many_are_left_in_the_wild_population/

18. http://news.bbc.co.uk/2/hi/science/nature/5085006.stm

19. http://www.britannica.com/EBchecked/topic/441032/giant-panda

20. http://www.thepetitionsite.com/takeaction/746/002/254/

21. http://wwf.panda.org/what_we_do/endangered_species/giant_panda/solutions/pandasuccess/

22. Ibid.

23. http://www.nrdc.org/globalwarming/fcons/fcons3.asp

24. http://news.mongabay.com/2009/0127-penguins.html

25. "March of the Penguins"

26. http://www.ipcc.ch/ipccreports/tar/wg2/index.php?idp=220

27. Saraux, C., et. al. " Reliability of flipper-banded penguins as indicators of climate change." Nature 469: 203-206. http://www.nipccreport.org/articles/2011/mar/29mar2011a6.html

28. Ibid.

29. D. Ksepka," Five Things You Never Knew About Penguins" http://blogs.scientificamerican.com/guest-blog/2010/12/20/5-things-you-never-knew-about-penguins/

30. http://news.bbc.co.uk/2/hi/sci/tech/3375447.stm

31. http://www.theaustralian.com.au/higher-education/frank-fenner-sees-no-hope-for-humans/story-e6frgcjx-1225880091722

32. http://www.bbcamerica.com/anglophenia/2011/09/royal-roundup-prince-charles-warns-of-human-extinction/

33. http://www.crh.noaa.gov/mkx/?n=taw-part10-usa_fatality_stats

34. Ibid.

35. http://www.newscientist.com/article/dn4259-european-heatwave-caused-35000-deaths.html

36. http://www.noaanews.noaa.gov/stories2011/20110309_russianheatwave.html

37. Bjorn Lomborg Cool It p.15

38. Martens et. al. "Climate Change- Mortality from Heat Stress http://www.ncbi.nlm.nih.gov/pubmed/9460815

39. http://paul.kedrosky.com/archives/2010/08/world_populatio.html
40. Al Gore An Inconvenient Truth
41. http://www.sciencedaily.com/releases/2008/10/081007073928.htm

42. Dr. Paul Reiter "Global Warming and Malaria: Knowing the Horse Before Hitching the Cart" www.malariajournal.com/article/10.1186/1475/2875/7/s1/s3

43. ibid
44. Ibid.
45. Ibid.
46. Ibid.
47. Ibid.

48. Soc. Sci. Med. 1992 Apr; 34(8):855-65. Yellow fever epidemics and mortality in the United States, 1693-1905. Patterson KD.
49. wwwnc.cdc.gov/eid/article/9/12/pdfs/03-0288.pdf
50. http://www.ncbi.nlm.nih.gov/pubmed/21028958
51. http://www.cdc.gov/ncidod/dvbid/westnile/birds&mammals.htm
52. Tamara Ben Ari, et. al. "Plague and Climate: Scales Matter" http://www.ncbi.nlm.nih.gov/pmc/articles/PMC3174245/
53. http://en.wikipedia.org/wiki/Lyme_disease
54. http://en.wikipedia.org/wiki/Cholera
55. email from Col. Ronald D. Harris M.D., USAF (from his experience in overseeing the smallpox vaccination program at Department of Defense MILVAX Agency.)

56. http://www.co2science.org/subject/g/summaries/disease.php

Chapter 8
1. 1 http://www.thefreedictionary.com/correlation
2. http://www.daviesand.com/Choices/Precautionary_Planning/New_Data/
3. Nicolas Caillon, et. al. "Timing of Atmospheric CO2 and Antarctic Temperature Changes Across Termination III" Science 14 March 2003: Vol. 299 no. 5613 pp. 1728-1731

Chapter 10
1. http://usatoday30.usatoday.com/weather/wheat7.htm
2. Ibid
3. http://usatoday30.usatoday.com/weather/wcstates.htm
4. http://www.ncdc.noaa.gov/oa/climate/globalextremes.html

Chapter 11 Misleading Charts

1. 1 [http://hadobs.metoffice.com/hadcrut3/diagnostics/index.html
Unfortunately no longer available on website.

Chapter 12
1. Lawrence Solomon The Deniers pp.13-15
2. http://www.appinsys.com/GlobalWarming/GW_Part2_GlobalTempMeasure.htm#historic
3. http://data.giss.nasa.gov/gistemp/station_data/
4. Ibid.
5. Let X represent the temperature for each of the cells outside Antarctica. To minimize variances and thereby minimize margin of error, we have stated that the temperature for each of those cells is the same. Since there are 1043 total cells of which 27 are in Antarctica, there are 1016 cells outside Antarctica. B29 in our table equals the sum of the temperatures outside Antarctica, i.e. 1016 X. We have as a given that the average global temperature is 150 C. Therefore, the sum total of temperatures in Antarctica and outside Antarctica divided by the total number of cells must equal 150C.
When we solve for X we get 15.750C. We then subtract 150 C (the given average global temperature) according to the formula shown above the table. We then square the difference and get 0.56. We then multiply by 1016, shown in D29 in our table. Adding D28 to D29 gives sum total of squares. Dividing that number (26555.66) by n-1(n= total number of cells) gives 25.49. Taking the square root gives 5.05. and then multiplying by 1.96 for a 95% confidence level gives 9.900C.
6. http://earthobservatory.nasa.gov/Features/WorldOfChange/decadaltemp.php
7. http://data.giss.nasa.gov/gistemp/station_data/
8. See chapter 16

Chapter 13
1. http://www.nasa.gov/topics/earth/features/rapid-change-feature.html
2. Bjorn Lomborg Cool It p 60
3. Lomborg Op. cit. p70 (Tol, R.S.J. "The Double Trade-off between Adaptation and Mitigation for Sea Level Rise")
4. Lomborg Op. cit. p 68

Chapter14
1. http://energyoptions-wind.com/EO_RPanels.html
2. http://rredc.nrel.gov/wind/pubs/atlas/tables/1-1T.html
3. http://en.wikipedia.org/wiki/Wind_turbine_design
4. http://www.decc.gov.uk/en/content/cms/meeting_energy/wind/onshore/questions/onshore_q2/onshore_q2.aspx
5. http://www.zerocarbonsolutions.com/index.php?page=46 19%Germany 22% Den-

mark Netherlands

6. http://www.offshorewind.biz/2012/12/21/ref-publishes-study-on-wind-turbine-lifespan-uk/life
7. http://en.openei.org/wiki/Wind_energy#Land_Requirements
8. http://en.wikipedia.org/wiki/Wind_turbine_design
9. Area= 63X63X3.14= 12,462.66 square meters
Power rating = 7,580,000 watts
Available power (25% load factor) = 7,580,000 watts X 0.25 = 1,895,000 watts
1,895,000 watts/12462.66 square meters= 152 watts/sq. meter
10. Alabama Power Miller Steam Plant Birmingham AL
11. http://en.wikipedia.org/wiki/Energy_in_the_United_States
12. Board of Equalization Jefferson County records. Estimate from aerial photography.
13. (126 X 5 X 23 X 126 X 5 X 23) + 4 X 126 = 209,960,604 sq. meters
209.960,604 / 625 = 335,937 sq. meters per E-126
14. " The Case against Windfarms" Dr. John Etherington (© Dr. JR Etherington). http://www.countryguardian.net/Case%20Introduction.htm
15. Ibid.
16. Ibid.
17. http://en.wikipedia.org/wiki/Ethanol_fermentation
18. http://www.epa.gov/cleanenergy/energy-and-you/affect/non-hydro.html biomass CO2
19. http://cen.acs.org/articles/89/web/2011/12/Burning-Biomass-Save-Money.html

Chapter15
1. http://www.ipcc-data.org/ddc_co2.html
2. http://www.ipcc.ch/publications_and_data/ar4/wg1/en/spmsspm-projections-of.html
3. Lomborg Op. cit. p.18
4. Lomborg Op. cit. p. 12
5. Lomborg Op. cit. p. 36
6. Ibid.
7. Lomborg Op. cit. pp. 43, 44
8. http://en.wikipedia.org/wiki/Copenhagen_Consensus

Chapter 16
1. S..Fred Singer "Global Warming: Man-Made or Natural" http://www.hillsdale.edu/news/imprimis/archive/issue.asp?year=2007&month=08
2. James Hansen Storms of my Grandchildren p. 12
3. John McLean "Prejudiced Authors, Prejudiced Findings" http://scienceandpublicpolicy.org/originals/prejudiced_authors_prejudiced_findings.html
4. Ibid.

5. Ibid.

6. Ibid.

7. Singer Op. cit.

8. http://news.heartland.org/newspaper-article/2008/07/01/30000-scientists-sign-petition-global-warming

9. Solomon Op. cit. p 193

10. Mark Morano http://www.climatedepot.com/a/9035/SPECIAL-REPORT-More-Than-1000-International-Scientists-Dissent-Over-ManMade-Global-Warming-Claims--Challenge-UN-IPCC--Gore

11. Christopher C. Horner Red Hot Lies p. 71

12. Horner Op. cit. p. 72

13. Horner Op. cit. pp. 72,73

14. Horner Op. cit. p. 72

15. Horner Op. cit. p. 113

16. Morano Op. cit.

17. Horner Op. cit. p. 257

18. Michael Crichton State of Fear (paperback) p. 629

19. http://www.canadafreepress.com/index.php/article/10567 0

20. http://www.eenews.net/public/Greenwire/2009/02/17/3

21. http://www.worldclimatereport.com/index.php/2012/07/24/illiteracy-at-nasa/

22. http://www.john-daly.com/quotes.htm

23. http://epw.senate.gov/hearing_statements.cfm?id=266543

24. http://climateaudit.org/2010/04/08/dealing-a-mortal-blow-to-the-mwp

25. Brian Sussman Climategate p.

Chapter 17

1. James Hansen "Global Warming: Twenty years later" http://www.huffingtonpost.com/dr-james-hansen/twenty-years-later-tippin_b_108766.html

2. Lawrence Solomon The Deniers pp. 161--167.

3. Fred Singer Unstoppable Global Warming: Every 1500 Years

4. IPCC Fourth Assessment Report:
Climate Change 2007: Working Group I: The Physical Science Basis 9.2.4 Summary<

5. Horner Op. cit. pp. 285-286

6. http://www.worldclimatereport.com/index.php/2007/11/28/terminating-warming-a-look-at-california/

7. http://antigreen.blogspot.com/2008/05/significant-warming-email-from-madhav.html

8. http://en.wikipedia.org/wiki/Antarctica

9. http://earthobservatory.nasa.gov/Features/WorldOfChange/sea_ice_south.php

10. http://wattsupwiththat.com/2012/09/10/icesat-data-shows-mass-gains-of-the-antarctic-ice-sheet-exceed-losses/

11. GEOPHYSICAL RESEARCH LETTERS, doi:10.1029/2012GL052559 "Increased ice

loading in the Antarctic Peninsula since the 1850s and its effect on Glacial Isotactic Adjustment"

12. http://www.appinsys.com/globalwarming/RS_EastAfrica.htm

13. http://www.ihatethemedia.com/12-more-glaciers-that-havent-heard-the-news-about-global-warming

14. http://www.biocab.org/carbon_dioxide_geological_timescale.html

15. http://www.purdue.edu/uns/x/2009a/090226HuberPete.html

Chapter18 Misleading Facts

1. http://news.mongabay.com/2009/0122-climate.html#

2. Ibid

3. James Hansen Storms of my Grandchildren preface

4. Ibid

5. Al Gore An Inconvenient Truth

6. Christopher C. Horner Red Hot Lies p 260

7. Hansen Op. cit.

8. http://marklynas.org

Chapter 19 Covering up facts

1. Op. cit. Horner p7

2. http://www.climatedepot.com/a/3212/Inconvenient-Questions-Stanford-U-Bans-Climate-Film-from-Airing-Interview-with-Cooling-turned-Warming-Prof-Stephen-Schneider--You-are-prohibited

3. Op. cit. Horner p 101

4. Op. cit. Horner p 127

5. Op. cit. Horner p.268

6. Ibid.

7. http://www.appinsys.com/GlobalWarming/UnprecedentedWarming.htm

Chapter 21 uncertainties in measurement

1. http://www.uoguelph.ca/~rmckitri/research/globaltemp/GlobTemp.JNET.pdf

2. Red Hot Lies p287, 288

3. http://wattsupwiththat.com/2009/05/03/climate-science-fraud-at-albany-university/

4. IPCC third assessment report scientific basis 2.2.2.1 Land-surface air temperature

Chapter 22 unknown uncertainties

1. http://climate.nasa,gov/uncertainties

INDEX

CPSIA information can be obtained at www.ICGtesting.com
Printed in the USA
LVOW02s0727170713

343239LV00001B/1/P